AF616649

INTERNATIONAL CENTRE FOR MECHANICAL SCIENCES

COURSES AND LECTURES - No. 140

LECTURES ON WIENER AND KALMAN FILTERING

T. KAILATH

DEPARTMENT OF ELECTRICAL ENGINEERING
STANFORD UNIVERSITY

SPRINGER - VERLAG

WIEN - NEW YORK

This is a new, revised edition of the book "Lectures on Linear Least-Squares Estimation" by the same author.

Printed in Italy

ISBN 3-211-81664-X Springer Verlag Wien-New York
ISBN 0-387-81664-X Springer Verlag New York-Wien

CONTENTS

LECTURES ON WIENER AND KALMAN FILTERING

by

Thomas KAILATH

Department of Electrical Engineering
Stanford University
Stanford, California 94305

PREFACE

These notes were first prepared for use in a special ten-hour lecture course at the CISM (Centre International des Sciences Mécaniques), Udine, Italy, June 11-22, 1972. The material in the first nine chapters was covered to varying degrees in the lectures, but advantage has been taken of the intervening time to slightly modify the text, especially with a view to including more up-to-date references. Though several interesting results have appeared since 1972, I have resisted the temptation to include them here, except for a partial outline in the concluding Chapter 10 and a few exercises added at various points. At the last minute, it was also decided to reprint a recent survey paper as an Appendix. Though no attempt has been made to prepare a comprehensive set of exercises, I may say that the notes in this form have been used as a text for a one-quarter course at Stanford.

I am grateful to many people for helping to deepen my understanding of the remarkably broad field of least-squares theory and I should like to mention W. Root, R. Price, D. Slepian, L. Zadeh, E. Parzen, F. Beutler, A. Balakrishnan, J. Clark, L. Shepp, R. Kalman, A. Shiryaev, R. Lip'ser, G. Kallianpur, P.A. Meyer, A. Yaglom, M. Zakai, J. Ziv, E. Wong, T. Kadota, T. Hida, H. Kunita, M. Hitsuda, H. Akaike, J. Rissanen, A. Schumitzky, B. Anderson, L. Silverman, A. Bryson, R. Bucy, L.A. Zachrisson, D. Lainiotis, J. Moore, G. Bierman, M. Davis, L. Ljung and several former students, especially J. Omura, P. Faurre, P. Frost, R. Geesey, T. Duncan, B. Gopinath, D. Duttweiler, H. Aasnaes, M. Gevers, H. Weinert, A. Segall, N. Krasner, M. Morf, G. Sidhu, B. Dickinson, B. Friedlander, A. Vieira, S.Y. Kung and B. Lévy. The manuscript was typed by Barbara McKee with the same skill, and the same cheerfulness and patience with last-minute changes and deadlines, as she has shown with essentially all my published (and unpublished) work. It also gives me great pleasure to acknowledge the consistent and generous research support of the Applied Mathematics Division of the Air Force Office of Scientific Research, which since 1963 has enabled me, with a minimum of overhead and reporting requirements, to pursue the studies that have provided the background on which these notes are based.

I am grateful to Professor Sandor Csibi of Budapest, Professor Giuseppe Longo of Trieste, and the CISM staff, for the invitation to present these lectures and especially for their patience with the long delay in my release of this still quite rough manuscript. The reasons for this behavior are well-known to all authors (and many more potential authors), but perhaps they were best put by a writer on a different subject: "What we do here is nothing to what we dream of doing" in Justine, by D.A.F. de Sade.

T. Kailath
Stanford, California
June 1975

SECOND EDITION

The increasing use of these short notes as a text for a first course in estimation theory led to a reprinting, of which advantage has been taken to correct several errors, update the references and make a few small changes, including a change in the title. A survey paper on discrete-time estimation has also been added as Appendix II.

T. Kailath
Stanford, California
September 1981

1. INTRODUCTION

Suppose we have two random variables X, Y with a known joint density function $f_{x,y}(.,.)$. Assume that in a particular experiment, the random variable Y can be measured and takes the value y. What can be said about the corresponding value, say x, of the unobservable variable X?

Suppose we make an estimate, say $\hat{x}$, of the value of X when Y = y, according to the rule

$$\hat{x} = h(y) ,$$

where x is the unobservable true value of X when Y = y, $h(\cdot)$ is some specified (linear or nonlinear) function. Let us denote

$$e = x - \hat{x} = x - h(y)$$
$$= \text{the error in our estimate.}$$

We can never hope to make e = 0 always (i.e., for every value of Y that may arise) because X and Y are chance (random) variables. Therefore, all we can hope for is to try to choose $h(\cdot)$ so as to minimize the expected (average) value of some function of the error, for example, the absolute value of the error, $|e|$, or the squared-error $|e|^2$, or some of the functions shown in Fig. 1.1.

The choice of an appropriate function of the error is obviously not a clear one and must depend upon the significance of different values of the error for the problem at hand – e.g., the function shown in Fig. 1.1(c) reflects the feeling that above a certain value "a miss is as good as a mile". With the squared-error function, small errors are weighted less than large errors and so on. Given this multitude of choices, it is somewhat remarkable that the squared-error criterion is in many ways the one deserving of the most study; the corresponding estimates will be called least-squares (l.s.) estimates, though other names are also used, especially minimum-mean-square-error (m.m.s.e.) estimates.

Among the many reasons for studying l.s. estimates, we might mention the following here:

i) the least-squares (l.s.) estimate can be specified explicitly as a conditional mean. Similar characterizations are not available for many of the error-functions shown in Fig. 1.1.

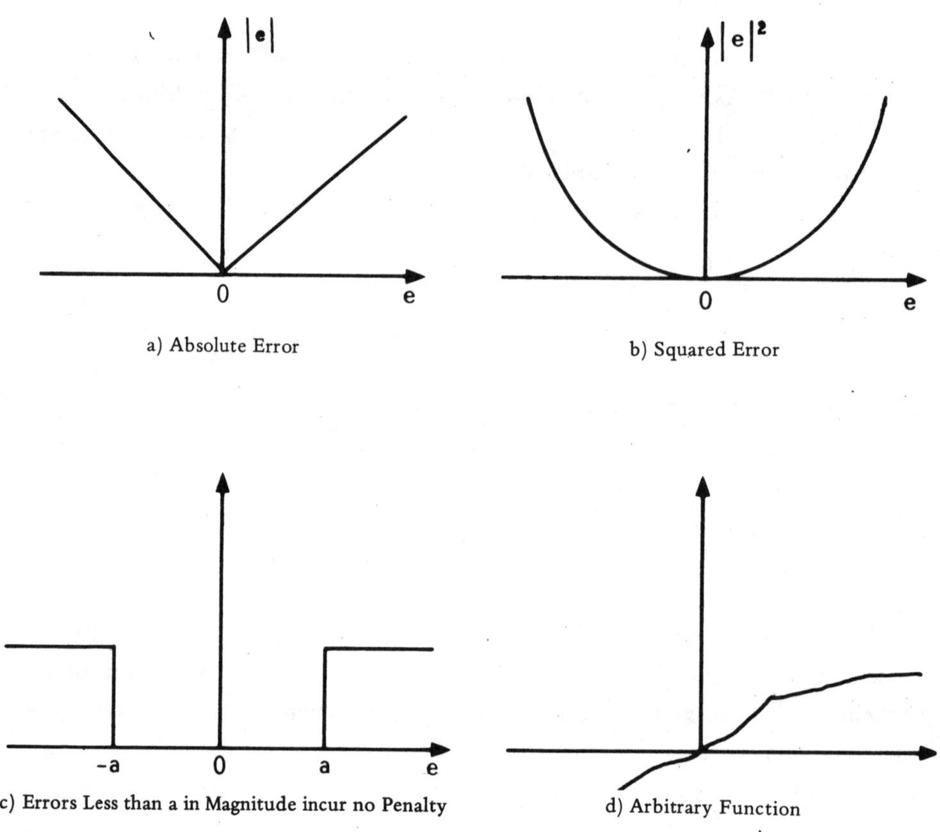

Fig. 1.1. Different Error Criteria

ii) For Gaussian random variables (a widely used model), the l.s. estimate is a linear function of the observables and thus has the general advantages of easier computation and implementation. Furthermore, for Gaussian random variables, the l.s. estimate turns out to essentially determine the optimum estimate for an arbitrary error function (cf. [1]).

iii) Suboptimum l.s. estimates are relatively easy to obtain. In particular, if the estimates are constrained to be linear functions of the observations, then it turns out that the estimator depends only upon the first and second-order statistics (i.e., means and covariances) of the random variables involved; moreover, the mean-square-error turns out to be independent of the actual observations and depends only on the prior statistical information on means and covariances. These important features bring great simplifications to the always-thorny questions of specification and determination of suitable (mathematical) models of physical problems.

iv) The solution of the stochastic linear least-squares estimation problem turns out to have many mutually illuminating connections with the ubiquitous deterministic problems of matrix inversion and the solution of linear (matrix and integral) equations.

v) The fact that the 1.s. estimate is a conditional expectation brings out certain connections with martingale theory, and thus provides useful results on the structure of likelihood ratios (useful in signal detection) and more generally on the structure of many stochastic processes.

Elaborations of some of these features will appear in the following chapters. A broader overview can be obtained from the survey papers [2], [3], and the collection [4]. (see also Appendices I and II).

REFERENCES

[1] V.S. Pugachev, "The Determination of an Optimal System by Some Arbitrary Criterion", **Automation and Remote Control,** vol. 19, pp. 513-532, June 1958. See also R.B. Street, **IEEE Trans. on Automatic Control,** vol. AC-8, pp. 375-376, October 1963. Also M. Zakai, **IEEE Trans. on Information Theory,** vol. IT-10, pp. 95-96, January 1964.

[2] T. Kailath, "The Innovations Approach to Detection and Estimation Theory", **Proc. IEEE,** vol. 58, pp. 680-695, May 1970.

[3] T. Kailath, "A View of Three Decades of Linear Filtering Theory", IEEE Transactions on Information Theory, vol. IT-20, no. 2, pp. 145-181, March 1974.

[4] T. Kailath, ed., "Linear Least-Squares Estimation", Benchmark Papers in Electrical Engineering, vol. 17. Academic Press, New York, 1977.

2. LEAST–SQUARES ESTIMATES – BASIC PROPERTIES

In this section we shall study a number of very simple l.s. estimation problems and examples, though we should emphasize that these simple discussions will bring out some of the most important concepts of the subject.

2.1. Least-Squares Estimates are Conditional Expectations

Returning to the simple problem posed in Section 1, we shall show that the least-squares estimate of a random variable X given another random variable Y is the conditional expectation of X given Y, written $E[X|Y]$.

Suppose that when $Y = y$, the true value of X is x, but that our estimate is $h(y)$. The error then is

$$x - h(y)$$

and this occurs with a probability

$$f_{X,Y}(x,y)\,dx\,dy$$

The mean-square-error (m.s.e.) therefore is

$$\text{m.s.e.} = \iint f_{X,Y}(x,y)\,[x - h(y)]^2\,dx\,dy$$

and this is to be minimized by proper choice of the function $h(.)$. Now recall the familiar decomposition

$$f_{X,Y}(x,y) = f_{X|Y}(x|y) f_Y(y)$$

where $f_{X|Y}(\cdot|\cdot)$ is the **conditional density function of X given Y** and $f_Y(.)$ is the (marginal) density of Y. Using this decomposition, we can rewrite the m.s.e. as

$$\text{m.s.e.} = \int f_Y(y)\,dy \left\{ \int f_{X|Y}(x|y)\,[x - h(y)]^2\,dx \right\}$$

Since a density $f_Y(.)$ is always nonnegative it is clear that the m.s.e. will be minimum if and only if the term in braces is a minimum. But it can be seen in many ways (e.g., by differentiation, or by noting that what we have is the moment of inertia of the "mass" distribution $f_{X|Y}(\cdot|y)$ about the point $h(y)$, or by other methods) that this will be so if and only if we choose

$$h(y) = \int x f_{X|Y}(x|y)\,dx$$

Now by definition, the right-hand side is the value at y of the conditional expectation of X given Y, E[X | Y]. Therefore, we can say that

$$h(Y) = E[X \mid Y].$$

The point is that since Y is a random variable we cannot say which value y it will assume in any particular occurence, and therefore the best estimate is also a random variable, which we shall denote $\hat{X}$,

$$\hat{X} = h(Y) = E[X \mid Y].$$

Example 2.1.1

Suppose X and Y are jointly Gaussian with

$$f_{X,Y}(x, y) = \frac{1}{2\pi(\sigma_x^2 \sigma_y^2 - \rho^2)^{\frac{1}{2}}} \exp - \frac{1}{2} \left\{ \frac{(x-m_x)^2}{\sigma_x^2 - (\rho^2/\sigma_y^2)} + \frac{(y-m_y)^2}{\sigma_y^2 - (\rho^2/\sigma_x^2)} - \frac{2\rho(x-m_x)(y-m_y)}{\sigma_x^2 \sigma_y^2 - \rho^2} \right\}$$

where

m_x = mean of X, m_y = mean of Y
σ_x^2 = variance of X, σ_y^2 = variance of Y
ρ = covariance of X and Y
= $E(X - m_x)(Y - m_y)$

Now it can be verified (using a table of integrals) that

$$E[X|Y] = m_x + \frac{\rho}{\sigma_y^2}(Y - m_y)$$

and that the minimum m.s.e. is

$$\text{m.m.s.e.} = \sigma_x^2 - (\rho^2/\sigma_y^2)$$

Though tedious, it will be useful for the student to actually carry out the algebra required to verify the above result.

Exercise 2.1.1.

Let

$$Y = X + V$$

where X and V are independent random variables, V is Gaussian with mean zero and variance unity, and X takes the values $\pm$ 1 with equal probability. Show that

$$\hat{X} = \tanh Y.$$

Exercise 2.1.2.

Let X and Y be jointly Gaussian zero-mean random variables and let

$$Z_1 = aY + b,$$

$$Z_2 = Y^2,$$

$$Z_3 = \cos Y.$$

Determine $E[X|Z_i]$, $i = 1,2,3$. What relation do they have to $E[X|Y]$?

Estimation Given Several Random Variables

Continuing with our discussion, suppose we wish to estimate the value of a random variable X given observations of two other random variables Y_1 and Y_2. As before, we can argue that the

$$\text{m.s.e.} = \int\int [x - h(y_1, y_2)]^2 f_{X,Y_1,Y_2}(x, y_1, y_2)\, dx\, dy_1\, dy_2$$

will be minimized by choosing

$$h(y_1, y_2) = \int x f_{X|Y_1,Y_2}(x|y_1, y_2)\, dx.$$

$$= E[X|Y_1 = y_1,\ Y_2 = y_2].$$

Therefore

$$\hat{X} = h(Y_1, Y_2) = E[X|Y_1, Y_2].$$

Furthermore it is clear that apart from a somewhat burdensome notation, the same arguments will show that the l.s. estimate of X given $\{Y_1, Y_2, \ldots, Y_n\}$ is

$$\hat{X} = E[X|Y_1, \ldots, Y_n].$$

Example 2.1.2. Jointly Gaussian Random Variables

Suppose $\{X, Y_1, \ldots, Y_n\}$ are jointly Gaussian with

$$EX = m_X,\ E\mathbf{Y} = E[Y_1, \ldots, Y_n]' = [m_{y_1}, \ldots, m_{y_n}]' = \mathbf{m_Y}$$

$$E(X - m_X)^2 = R_{xx}$$

$$E[\mathbf{Y} - \mathbf{m}_Y][\mathbf{Y} - \mathbf{m}_Y]' = \mathbf{R}_{YY}, \text{ a nonsingular matrix}$$

$$E[X - m_X][\mathbf{Y} - \mathbf{m}_Y]' = R_{XY} = R'_{YX}.$$

Here the primes denote transposes and we are using an obvious matrix notation. Now it can be verified that

$$\hat{X} = E[X | Y_1, \ldots, Y_n]$$
$$= m_X + R_{XY} R_{YY}^{-1} [\mathbf{Y} - \mathbf{m}_Y].$$

with the corresponding minimum mean square error

$$\text{m.m.s.e.} = R_{xx} - R_{XY} R_{YY}^{-1} R_{YX}.$$

where we have written R_{xx} for σ_x^2.

Exercise 2.1.3.

Derive the results stated in Example 2.1.2. [This can be done in several ways, but it will be also a useful review of some matrix theory to proceed via diagonalization of R_{yy}.]

Exercise 2.1.4.

Let X, Y_1, Y_2 be independent Gaussian zero-mean, unit-variance random variables and let $R = \sqrt{Y_1^2 + Y_2^2}$. Determine $E[X| R]$ and $E[Y_1 | R]$.

Estimation Given a Random Process

The next step in generalization is to consider the l.s. estimate of a random variable X given observations of a section of a stochastic process $\{Y(\tau), a \leqslant \tau \leqslant b\}$. We might reasonably expect, by an extension of the previous arguments, that

$$\hat{X} = E[X | \{Y(\tau), a \leqslant \tau \leqslant b\}].$$

This is correct, but only if the right-hand-side is properly defined (computed) ! The difficulty, as the reader will immediately recognize, lies in trying to compute (define) the "conditional density functional"

$$f(X|\{Y(\tau), a \leqslant \tau \leqslant b\}).$$

A first guess might be the following: let us assume that the random (or stochastic, the two words will be used interchangeably) process $\{Y(\tau), a \leqslant \tau \leqslant b\}$ is "smooth" enough that it can be represented by a countable number of random variables $\{Y_i, i = 1,2, \ldots, \infty\}$, e.g., generalized "Fourier" coefficients with respect to some countable basis of orthonormal functions. Then it is reasonable to define

$$f(X|\{Y(\tau)\}) = \lim_{n \to \infty} f(X|\{Y_1, \ldots, Y_n\}).$$

Unfortunately it often turns out that the limit of the (well-defined) quantity on the right is not generally well behaved – it may not exist or it may be infinite.

The above approach can be suitably modified so as to work, by using the notion of "generalized (Radon-Nikodym) derivatives", but here we shall pursue an alternative and more immediately useful procedure: we shall use a definition of conditional expectation that does not rely on the notion of conditional density.

We shall use the definition (justified in many books on probability theory) that:

"The conditional expectation of a random variable X given a stochastic process $\{Y(.) \equiv \{Y(\tau), a \leqslant \tau \leqslant b\}\}$ (the cases of a random variable Y, or a random vector $\mathbf{Y}' = [Y_1, \ldots, Y_n]'$, are subsumed by this) is the unique random variable that

i) is a functional of Y(.), say $h(\{Y(\tau), a \leqslant \tau \leqslant b\})$

ii) satisfies the "orthogonality condition"

$$E[X - h(\{Y(\tau), a \leqslant \tau \leqslant b\})]\, g(\{Y(\tau), a \leqslant \tau \leqslant b\}) = 0,$$

for all functionals g(.) (that are random variables for which the above expected value is meaningful). Such a functional h(.) will be called the conditional expectation of X given Y(.) and will be written $E[X|Y(.)]$ or $E[X|\{Y(\tau), a \leqslant \tau \leqslant b\}]$.

It is shown in many textbooks that this "descriptive" definition coincides with the usual "constructive" definition, whenever the latter is meaningful. We shall only provide the following simple example.

Example 2.1.3. Equivalence of two definitions

Let X and Y be random variables with a conditional density function

$f_{X|Y}(\cdot\,|\,\cdot)$ so that we can define

$$E[X|Y] = \int x f_{X|Y}(x|Y)\ dx$$

Then we can show that

$$E[[X - E[X|Y]]g(Y)] = 0, \quad \text{all} \ \ g(Y)$$

or equivalently that

$$E[E[X|Y]g(Y)] = E[Xg(Y)], \quad \text{all} \ \ g(Y).$$

To do this, note that

$$\begin{aligned} \text{the LHS} &= \int\int dx\ dy\ f_{X,Y}(x,y)g(y) \int u f_{X|Y}(u|y)\ du \\ &= \int dy\ \ g(y)f_Y(y) \int u f_{X|Y}(u|y)\ du \\ &= \int\int du\ \ dy u f_{X,Y}(u,y)g(y) \\ &= \text{the RHS.} \end{aligned}$$

Conversely, suppose h(Y) is a function of Y such that

$$E[Xg(Y)] - E[h(Y)g(Y)], \quad \text{all} \ \ g(Y).$$

Then

$$\begin{aligned} \text{LHS} &= \int\int xg(y)f_{X,Y}(x,y)\ dx\ dy \\ &= \int g(y)f_Y(y)\ dy \int x f_{X|Y}(x|y)\ dx \end{aligned}$$

Comparing with the RHS

$$\begin{aligned} \text{RHS} &= \int\int h(y)g(y)f_{X,Y}(x,y)\ dx\ dy \\ &= \int g(y)h(y)f_Y(y)\ dy \end{aligned}$$

shows that we must have

$$h(y) = \int x f_{X|Y}(x|y)\ dx\,,$$

wherever $f_Y(y) > 0$ (and we do not care what values are assumed when $f_Y(y) = 0$, for such values make no difference to the integrals).

Least-Squares Estimation and Conditional Expectations

We shall now show how the new definition of conditional expectation gives a simple approach to the least-squares problem; as the reader will note, the proof is in fact even simpler, at least notationally, than the previously given proof for a finite number of random variables.

Given X and an observed process Y(.), suppose we wish to find a functional h(Y(.)) such that

$$E[X - h(Y(.))]^2 = \text{minimum}.$$

Now note that we can write

$$\begin{aligned} E[X - h(Y(.)\,]^2 &= E[X - E[X|Y(.)] + E[X|Y(.)] - h(Y(.)\,]^2 \\ &= E[X - E[X|Y(.)]]^2 + E[E[X|Y(.)] - h(Y(.)\,]^2 \\ &\quad + 2E[[X - E[X|Y(.)\,]][E[X|Y(.)] - h(Y(.)\,]]. \end{aligned}$$

But the last term is zero by the defining property (ii) of the conditional expectation and now it is clear that the sum of the two squared-terms will be minimum if and only if

$$h(Y(.)) = E\,[X\,|\,Y(.)], \text{ the conditional expectation of X given Y(.).}$$

This is certainly a simple derivation and a nice result, but the problem is of course that we have not shown how the conditional expectation may actually be obtained in any particular problem. Unfortunately there is only a small handful of cases for which explicit analytical solutions can be obtained, or even computationally feasible approximation procedures are known. Therefore suboptimum estimates of various kinds are sought, with the most common estimates that are constrained to being linear functionals of the observations. The benefits of such a constraint will be made clear in the next section, but we might note here that as shown by Examples 2.1.1 and 2.1.2, when all quantities are jointly Gaussian, the least-squares estimate is automatically linear. Since Gaussian models are so widely used, this is a strong reason for interest in linear l.s. estimates.

2.2. Linear Least-Squares Estimates

When we constrain the estimates to be linear, the previous results do not apply (at least directly) and we have to begin again. It will be useful, as before, to start with the simplest example and build up to the general case.

Thus suppose we wish to estimate the value of a random variable X by means of some linear operation on the value of an observed random variable Y, say as

$$\hat{X} = hY + g$$

where $\{h, g\}$ are to be chosen to minimize the mean-square-error

$$E[X - hY - g]^2 = E X^2 + h^2 E Y^2 + g^2 - 2gm_x + 2hgm_y - 2hEXY$$

where m_x and m_y are the means of X and Y respectively.

Important Note: We use the same symbol $\hat{X}$ for both the (constrained) linear l.s.e. and the true (unconstrained) l.s.e. in order to keep the notational burden down – it will not be hard to see from the context which type of estimate is being considered.

Returning to our problem, it is clear that differentiation with respect to h and g yields the following equations for a minimum,

$$g = m_x - h\, m_y$$
$$(EY^2)h = (EXY - g\, m_y)$$

Solving these gives us

$$h = \rho/\sigma_y^2\,,\; g = m_x - \rho m_y/\sigma_y^2\,,$$

where

$$\rho = E(X - m_x)(Y - m_y) = EXY - m_x m_y$$
$$\sigma_y^2 = E(Y - m_y)^2 = EY^2 - m_y^2\,.$$

Therefore the **linear** least-squares estimate of X given Y is

$$\hat{X} = m_x + (\rho/\sigma_y^2)(Y - m_y)$$

and the corresponding mean square error is

$$\text{m.s.e.} = \sigma_x^2 - (\rho^2/\sigma_y^2)\,.$$

Comparing these answers with the results of Example 2.1. shows that, as expected, the linear 1.s. estimate coincides with the (unconstrained) 1.s. when the random variables X and Y are jointly Gaussian. Another perhaps more important remark is that determination of the linear 1.s.e. requires only knowledge of the means and covariances of the random variables involved and **does not** require knowledge of the joint density function as would be needed for the true 1.s.e. (the conditional mean). This fact is a tremendous simplification in physical problems, where means and covariances are generally very much easier to determine than joint density functions. As we shall see these remarks carry over even to more complicated linear 1.s. estimation problems.

Thus suppose we wish to estimate the value of X as a linear combination of the values of n other random variables $\{Y_1, \ldots, Y_n\}$, so that

$$\hat{X} = \mathbf{h}\,\mathbf{Y} + g$$

where $\mathbf{h} = [h_1, \ldots, h_n]$ and g are again to be chosen so that

$$E[X - h\,Y - g]^2 = \text{minimum}$$

Now straightforward but notationally more cumbersome algebra shows that

$$h\ R_{YY} = R_{XY},\ g = m_X - h\ m_Y\,,$$

where

$$R_{YY} = E[Y - m_Y]\,[Y - m_Y]' = \text{the covariance matrix of the random vector } \mathbf{Y}$$

and

$$R_{XY} \doteq E\,X\,\mathbf{Y}' = R'_{YX}\,.$$

Therefore

$$\hat{X} = m_x + R_{XY}R_{YY}^{-1}\,[\mathbf{Y} - \mathbf{m}_Y]\,.$$

Moreover the corresponding m.s.e. is

$$\text{m.s.e.} = R_{XX} - R_{XY}R_{YY}^{-1}R_{YX}$$

Again we note that this solution only requires knowledge of "first-and second-order statistics" (i.e., the means and covariances), and that it coincides with the true least-squares estimate when the $\{X, Y_1, \ldots, Y_n\}$ are jointly Gaussian. Note also that the mean-square error is independent of the actual values of the random variables $\{X, Y_1, \ldots, Y_n\}$.

Before proceeding to consider estimation given a random process let us note a simple device for reducing the notational burden in calculations like those above.

Reduction to the Zero-Mean Case

The algebra in the above calculations can be considerably reduced by assuming that the random variables all have zero mean-values. To take only the simplest case of estimating X as hY+g, note that in the zero-mean case, we immediately have

$$E[X - hY - g]^2 = \sigma_X^2 + h^2 \sigma_Y^2 + g^2 - 2h\rho,$$

and the optimum h and g are readily found as

$$g = 0, \quad h = \rho/\sigma_Y^2 .$$

The reader should verify that even greater savings accrue in the problem of estimating X from vector observations.

Moreover, we see that the above algebraic simplifications will ensue automatically even in the general (nonzero means) case, if we first assume that the random variables are "centered", namely X is replaced by $X - m_X$ and **Y** by $Y - m_Y$. **Therefore we shall henceforth assume, unless explicitly stated otherwise, that all random variables arising in the problem have mean zero.**

We return now to the following linear estimation problem.

Linear Estimation Given a Random Process

Suppose we wish to find the 1.1.s. estimate of a random variable X given observations of a random process $\{Y(\tau), a \leq \tau \leq b\}$. We shall assume that the most general linear functional of Y will have the form

$$\int_a^b h(\tau) Y(\tau) d\tau$$

Then the problem of choosing h(.) so as to minimize the mean-square-error will reduce to an "infinite-dimensional" minimization problem i.e., to a problem in the calculus of variations. This can be done, but here we shall take a simpler route, based on first approximating the infinite-dimensional problem by a finite-dimensional problem and then taking the limit. (The reader may object that we rejected this approach in Sec. 2.1; however, here the difference is that because of the constraint of linearity, we shall not need to consider infinite-dimensional conditional density functions.)

To carry through this approach, let us first note that a smooth random process $\{Y(\tau), a \leqslant \tau \leqslant b\}$ can be approximately represented as (cf. Fig. 2.1)

$$Y(\tau) \doteq \sum_{i=0}^{n-1} Y(\tau_i)\sqrt{\Delta}\, p(\tau - i\Delta) = \sum_{i=0}^{n-1} Y_i\, p(\tau - i\Delta), \text{ say}$$

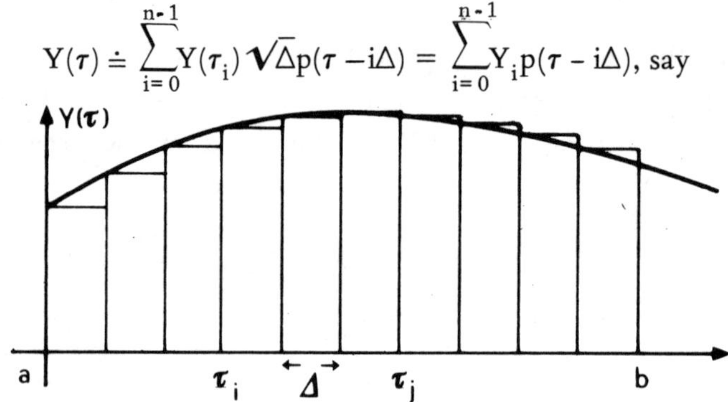

Fig. 2.1. Approximation of a Smooth Random Process.

where

$$p(\tau) = \begin{cases} \frac{1}{\sqrt{\Delta}}, & 0 \leqslant \tau \leqslant \Delta \\ 0, & \text{elsewhere.} \end{cases}$$

Then the 1.1.s.e. of a random variable X given $\{Y(\tau), a \leqslant \tau \leqslant b\}$ can be approximately obtained (by virtue of the result of the previous finite-dimensional problem) as

$$\hat{X} \doteq \mathbf{h}\ \mathbf{Y}, \text{ where } Y' = [Y_0, \ldots, Y_{n-1}]$$

and

$$\mathbf{h}\ \mathbf{R}_{YY} = \mathbf{R}_{XY}\ .$$

More explicitly (recall our assumption of zero means)

$$\sum_j h_j (EY_j Y_i) = E\, X\, Y_i\,, \quad i = 0, \ldots, n-1$$

or

$$\sum_j h_j R_{YY}(\tau_j,\tau_i)\Delta = R_{XY}(\tau_i)\sqrt{\Delta}, \quad \tau_i = i\Delta$$

Now let h(.) be a function approximately represented as

$$h(\tau) \doteq \sum_0^{n-1} h(\tau_i)\sqrt{\Delta}p(\tau - i\Delta), \qquad a \leqslant \tau \leqslant b$$

$$= \sum_0^{n-1} h_i p(\tau - i\Delta), \text{ say }.$$

Then the above equation is

$$\sum_j h(\tau_j) R_{YY}(\tau_j, \tau_i)\Delta = R_{XY}(\tau_i), i = 0, \ldots, n-1.$$

Finally let

$$\tau_i = \text{a fixed point t in } [a, b]$$

and consider the limit as $\Delta \longrightarrow 0$. Actually, the $\{\tau_j\}$ have to be written as $\{\tau_j^{(n)}\}$ to denote say the n-th partition of [a, b], but the reader can readily be persuaded that in the limit we get the following equation for h(.),

$$\int_a^b h(\tau)R_{YY}(\tau, t)d\tau = R_{XY}(t), \quad a \leqslant t \leqslant b.$$

This is known as an integral equation because the unknown function appears under an integral sign. Note also that

$$\hat{X} \doteq \mathbf{h}\ \mathbf{Y} = \sum_0^{n-1} h_i Y_i = \sum_0^{n-1} h(\tau_i)\, Y(\tau_i)\Delta$$

becomes, in the limit as $\Delta \to 0$, $n \to \infty$

$$\hat{X} = \int_a^b h(\tau)Y(\tau)d\tau.$$

Finally, the mean-square-error associated with this estimate is

$$\text{m.s.e.} = R_{xx} - \int_a^b \int_a^b h(\tau)h(\sigma)R_{YY}(\tau, \sigma)d\tau d\sigma.$$

The knowledgeable reader can verify that these will indeed be the results obtained via a calculus of variations argument. Of course, no matter how they are obtained, we are still left with the problem of solving the integral equation. Explicit solutions can only be obtained in special (but still widely useful) cases, and these will be the main object of attention in later chapters.

It might be noted that a not unsatisfactory approximate solution can always be obtained by going to the approximate estimation problem, where the solution of the integral equation is replaced by the more conventional problem of inverting a matrix. This is in fact a feasible solution, though certain caveats must be noted. First, in general a more sophisticated "discretization" scheme would be used, e.g., Gaussian quadrature. Secondly, even after suitable discretization, we will be faced with the problem of inverting a matrix of possibly large dimensions. Since the effort required to invert an n x n matrix is of the order of n^3 computations, we soon see that even with the largest computers, some attention must be paid to exploiting possible special structures in the matrices when n is large (e.g., n = 1000).

In fact, the rewards of trying to determine suitable structures have been spectacular, since many physical problems certainly do have special characteristics. This is the reason for the still-growing interest in the linear estimation problem and it will be the justification for our rather close examination in later chapters of the properties of special linear least-squares estimation problems. For reasons of space, among others, we shall not in these notes go into any actual applications.

The following example illustrates well some of the above comments about the numerical solution of integral equations and the importance of a priori structural information.

Example 2.2.1

Let Y(.) be a zero-mean stationary random process with covariance function

$$R_{YY}(t,\tau) = \exp - \alpha\,|t-\tau|\,.$$

Let a = 0, b = T, and X = Y(T + λ), λ > 0. Then the optimum weighting function h(.) is the solution of the integral equation

$$\int_0^T h(\tau)e^{-\alpha|t-\tau|}d\tau = e^{-\alpha(T+\lambda-t)}, \quad 0 \leqslant t \leqslant T\,.$$

It can be seen that solution depends upon T and can be written

$$h_T(\tau) = \delta(\tau - T)e^{-\alpha\lambda}, \; \delta(\,.\,) = \text{Dirac delta function}$$

so that

$$\hat{X} = \int_0^T h(\tau)Y(\tau)d\tau = e^{-\alpha\lambda}Y(T)\,.$$

In other words, the 1.1.s.e. of $Y(T+\lambda)$ given observations of $Y(t)$, $0 \leqslant t \leqslant T$, is an exponentially damped multiple of the most recent value. This is a striking result, which should have a nice significance and should have been predictable. This point will be pursued later, but here we would like to note that the solution h(.) of the integral equation can be very "nonsmooth", so that a very "fine" discretization (and therefore a very large n) would be needed to capture the true nature of the solution. In fact, unless we knew something about what to expect in the solution, it would be almost impossible to discover it by plunging blindly ahead to "solve" a discrete-time approximation on a computer.

Many other examples can be given to illustrate the necessity of gaining insight into the kinds of solutions that can be expected, even if ultimately the actual solution will have to be carried out on a computer. Therefore a close study of key special cases is vital and will be undertaken in later chapters. However before doing so, it will be useful to give a geometric interpretation of the least-squares estimation problem. This interpretation will enable us to reduce the conceptual and algebraic burden in many long chains of argument in later chapters.

Exercise 2.2.1.

Show that the best estimate of X given the criterion of minimum absolute error $E\,|X-\hat{X}|$ is the conditional median value of X given the observations.

HINT:

$$E|X-m| = \int_{-\infty}^{m} (m-u)p_x(u)\; du + \int_{m}^{\infty} (u-m)p_x(u)\; du$$

Exercise 2.2.2.

Let X and Y be jointly (zero mean) Gaussian random variables. Show that $\hat{X} \triangleq E[X|Y]$ not only minimizes the "total" mean squared error $E_{XY}(X-\hat{X})^2$ but that it also minimizes $E_{X|Y}(X-\hat{X})^2$, the "conditional" mean squared error, and that in fact the conditional m.s.e. is the same as the overall m.s.e. This property is quite striking because it says no matter how "bad" a particular piece of observed data may appear to be relative to some other piece of data, the mean-square error is independent of the actual data. This is an important property of the Gaussian case.

Show that it also holds whenever $E[X|Y]$ is a linear function of Y.

Exercise 2.2.3.

Show that $\hat{X} = R_{XY}R_Y^{-1}Y$ is also the least squares estimate of X among the "randomized" rules, i.e., if we let $\hat{X}^* = R_{XY}R_Y^{-1} Y + V$, where V is a random variable independent of X and Y, then $E(X - \hat{X}^*)^2$ is no less than the error obtained with $\hat{X} = R_{XY}R_Y^{-1} Y$ [i.e., adding an independent "dithering" noise to the estimate does not help !] .

2.3. Geometric Interpretations and the Orthogonality Conditions

It will be helpful to begin with certain familiar nonstatistical linear least-squares approximation problems that have an obvious geometric interpretation. These will lead us to the concept of an abstract normed space and from there to a geometric view of least-squares estimation.

Approximation in 3-Space

Suppose we have 2 vectors $\vec{Y}_1, \vec{Y}_2$ in ordinary Euclidean three--dimensional space and wish to form a linear combination of them that will best approximate another vector X in the sense that

$$\|\vec{X} - h_1\vec{Y}_1 - h_2\vec{Y}_2\|^2 = \text{minimum}$$

where

$$\|\vec{A}\|^2 = \text{the squared-length of } \vec{A}$$

It is clear by differentiation, for example, that the optimum $\{h_i\}$ are the solution of the equations

$$\begin{bmatrix} \langle\vec{Y}_1, \vec{Y}_1\rangle & \langle\vec{Y}_1, \vec{Y}_2\rangle \\ \langle\vec{Y}_2, \vec{Y}_1\rangle & \langle\vec{Y}_2, \vec{Y}_2\rangle \end{bmatrix} \begin{bmatrix} h_1 \\ h_2 \end{bmatrix} = \begin{bmatrix} \langle\vec{X}, \vec{Y}_1\rangle \\ \langle\vec{X}, \vec{Y}_2\rangle \end{bmatrix}$$

where

$$\langle\vec{A}, \vec{B}\rangle = \text{the dot-product of } \vec{A} \text{ and } \vec{B}.$$

However it is no doubt also obvious to the reader that the above equation can be somewhat more elegantly obtained by the following geometric argument. The vectors $\vec{Y}_1$ and $\vec{Y}_2$ define a plane in which any linear combination, $h_1\vec{Y}_1 + h_2\vec{Y}_2$, lies. Now the combination that minimizes the squared-length of the error is clearly obtained by dropping a perpendicular from the vector $\vec{X}$ to the plane of the vectors $\{Y_1, Y_2\}$. Or equivalently, the vector $\vec{X}_0$ corresponding to the optimum combination is uniquely determined by the fact that

$$\vec{X} - \vec{X}_0 \perp \text{the plane of } \{\vec{Y}_1, \vec{Y}_2\},$$

which in turn is clearly equivalent to the conditions

$$\vec{X} - \vec{X}_0 \perp \vec{Y}_i, \quad i = 1, 2,$$

where of course $\perp$ stands for "is orthogonal to". If $\vec{X}_0 = h_1 \vec{Y}_2 + h_2 \vec{Y}_2$, these orthogonality conditions give us the equations

$$\langle \vec{X}, \vec{Y}_1 \rangle = h_1 \langle \vec{Y}_1, \vec{Y}_1 \rangle + h_2 \langle \vec{Y}_2, \vec{Y}_1 \rangle$$

and

$$\langle \vec{X}, \vec{Y}_2 \rangle = h_1 \langle \vec{Y}_1, \vec{Y}_2 \rangle + h_2 \langle \vec{Y}_2, \vec{Y}_2 \rangle,$$

which are of course the same as stated above.

Extension to Higher Dimensional Space

Now suppose we have vectors in some abstract n-dimensional space and that we have a notion of inner product $\langle , \rangle$ and of length, $\|\cdot\|^2 = \langle \cdot, \cdot \rangle$, in this space.

Now suppose we wish to find $\{h_1, \ldots, h_n\}$ so that

$$\vec{X}_0 = \sum_1^n h_i \vec{Y}_i$$

forms a best approximation to a given n-vector $\vec{X}$ in the sense that

$$\|\vec{X} - \vec{X}_0\| = \text{minimum}.$$

Extending our geometric intuition from 3-space, we would say that the optimal $\vec{X}_0$ would be the vector determined by the projection of $\vec{X}$ on the (hyper-) plane determined by the $\{\vec{Y}_i\}$. In other words, we would say that the optimum $\{h_i\}$ would be uniquely determined by the **orthogonality conditions**

$$\vec{X} - \sum_1^n h_i \vec{Y}_i \perp \vec{Y}_j, \quad j = 1, \ldots, n,$$

which lead to the matrix equation (in an obvious notation)

$$\left[\langle \vec{Y}_i, \vec{Y}_j \rangle \right]_{i,j=1}^n \mathbf{h} = \left[\langle \vec{X}, \vec{Y}_i \rangle \right]_{i=1}^n$$

[Note of course that $\vec{A} \perp \vec{B}$ denotes $\langle \vec{A}, \vec{B} \rangle = 0$]. This seems very reasonable and reduces nicely to the 3-space problem. But one may be bothered by the fact that there are (nonintuitive) phenomena in n-space. Therefore, how can one be sure that the geometrical arguments just used to find X_0 are as valid in n-space as they "obviously" are in 3-space, especially as we have "abstract" definitions of length and angle (inner-product) in n-space ? The answer is that our geometrical arguments will be valid if the definition used for the "inner-product" $\langle \cdot, \cdot \rangle$ satisfies three requirements:

a) linearity : $\langle a_1 \vec{X}_1 + a_2 \vec{X}_2, Y \rangle = a_1 < \vec{X}_1, \vec{Y}_1 > + \; a_2 < \vec{X}_2, \vec{Y}_2 >$

b) symmetry: $\langle \vec{X}, \vec{Y} \rangle = \langle \vec{Y}, \vec{X} \rangle$

c) nondegeneracy: $\|\vec{X}\|^2 = \langle \vec{X}, \vec{X} \rangle$ is strictly positive unless $\vec{X} \equiv 0$.

For simplicity, we have assumed here that the vectors are real-valued; this will suffice for all our applications. With $\vec{X} \sim [x_1, \ldots, x_n]'$, $\vec{Y} \sim [y_1, \ldots, y_n]'$, we can verify that the following common definitions for $\langle \cdot, \cdot \rangle$, in n-space do satisfy the above requirements:

i) $\langle \vec{X}, \vec{Y} \rangle_1 = \sum_1^n x_i y_i$

ii) $\langle \vec{X}, \vec{Y} \rangle_2 = \sum_1^n x_i y_i \lambda_i , \quad \lambda_i > 0$

iii) $\langle \vec{X}, \vec{Y} \rangle_3 = \sum_i \sum_j a_{ij} x_i y_j , \; A = [a_{ij}]$ strictly positive-definite.

Any other definition would be equally permissible if it satisfied the three conditions of linearity, symmetry and nondegeneracy. We shall see the point of this well in the next problem.

Approximation in a Space of Functions

In Fourier analysis, the following is a common problem: given a set of continuous time-functions $Y_i(t)$, $a \leqslant t \leqslant b$, $i = 1, \ldots, n$ find the linear combination $\sum_1^n h_i Y_i(.)$ that will best approximate a given function $X(.)$ in the sense that

$$\int_a^b [X(\tau) - \Sigma h_i Y_i(\tau)]^2 \, d\tau = \text{minimum}.$$

Again it is easy to see by direct calculation that the optimum h can be found as the solution of the following matrix equation (in an obvious notation)

$$\left[\int_a^b Y_i(t) Y_j(t) \, dt \right]_{i,j=1}^n \mathbf{h} = \left[\int_a^b X(t) Y_i(t) \, dt \right]_{i=1}^n$$

However it is perhaps less well known to the reader that the above problem and solution admit of the following geometric interpretation. Consider an infinite-dimensional space with one "coordinate axis" for each t in $[a, b]$. Then any function Y(.) can be regarded as a "vector" in this space, defined by having component Y(t) along the coordinate axis corresponding to t. The inner-product of two vectors in this space can be defined as

$$\langle Y_1(\cdot), Y_2(\cdot)\rangle = \int_a^b Y_1(t)Y_2(t)\ dt$$

and it can be verified that this definition meets the three requirements of linearity, symmetry and nondegeneracy noted earlier. Therefore our geometrical viewpoint will be valid and we can state that the optimal $\{h_i\}$ will be uniquely determined by the conditions

$$X(\cdot) - \sum_1^n h_i Y_i(\cdot) \perp Y_j(\cdot), j = 1, \ldots, n$$

which can easily be seen to be equivalent to the matrix equations given earlier.

Many other function spaces and inner-products could be similarly used. For example, we could consider a space of differentiable functions with inner product

$$\langle Y_1(\cdot), Y_2(\cdot)\rangle_2 = \int_a^b \dot{Y}_1(t)\dot{Y}_2(t)\ dt + Y_1(a)Y_2(a)$$

and exactly the same "orthogonality conditions" would suffice to determine the minimum-error solution.

Clearly an advantage of the geometric interpretation is that the "same" solution can be applied to a large variety of apparently different problems. As a final example, we shall show how to apply this geometric picture to the statistical least-squares problems discussed in Section 2.2.

Geometric Interpretations for Random Variables

It suffices to say that we can think of (zero-mean) random variables X,Y,Z,. . . . as vectors in some abstract space with an inner product defined as

$$\langle X, Y\rangle = EXY\ .$$

It can readily be verified that this is a legitimate inner-product because it has the properties of linearity, symmetry and nondegeneracy.

Therefore, for example, to find $\{h_i\}$ so that

$$E\left[X-\sum_1^n h_iY_i\right]^2 = \|X-\sum_1^n h_iY_i\|^2 = \text{minimum}$$

it suffices to note that we must have

$$X-\sum_1^n h_iY_i \perp Y_j, \quad j=1,\ldots,n$$

and it will be seen that this of course leads to the same equations as in Section 2.2.

However the real rewards of the geometric interpretation arise in the problem of estimating X from observations of a random process Y(.). If we think of the random process as an (indexed) collection of random variables $\{Y(\tau), a\leqslant\tau\leqslant b\}$, then the optimum estimate

$$\hat{X}=\int_a^b h(\tau)\,Y(\tau)\,d\tau$$

will be uniquely defined by the orthogonality conditions

$$X-\int_a^b h(\tau)Y(\tau)\,d\tau \perp Y(t), \quad a\leqslant t\leqslant b,$$

which are equivalent to

$$EXY(t)=\int_a^b h(\tau)EY(\tau)Y(t)\,d\tau$$

or

$$R_{XY}(t)=\int_a^b h(\tau)R_{YY}(t,\tau)\,d\tau,$$

an equation that was obtained much less directly (by a limiting argument) in Section 2.2.

We shall see several other examples of such applications of the orthogonality conditions in later sections. However, while emphasizing the benefits to be gained from the geometric viewpoint, we should state that the justifications for this viewpoint that we presented above are really more a "plausibility argument" than a rigorous proof. This cautionary remark, which actually applies to essentially all the "proofs" presented in these notes, may make some readers feel extremely insecure. Since I feel that in a first course (which for many people is often also a last course), it is more important to emphasize intuition and interpretation, the only reassurance I can offer any reader is that all the arguments presented in these notes can be rigorized. The amount of background knowledge and of effort required to do this varies with the results, but I have found that in all cases the "heuristic" argument can be made the basis of a rigorous proof.

As our final topic in this subsection however, we make a cautionary remark and also attempt to show the relevance of one of the basic points of probability theory, which however is very rarely encountered in practice, namely the significance of the formal definition of random variable.

A Caution

We often think of a random process as a collection of (sample) time functions and in view of our discussion above of a geometric viewpoint for time-function, we might be tempted to seek a geometric interpretation for least-squares estimation in terms of time functions. However, as a little reflection will show, there is no way of getting a satisfactory geometric picture along these lines and we must think of the random process as a collection of random variables.

When it is necessary to emphasize this aspect, we might write the random process as $\{Y(t,\omega),\ a \leqslant t \leqslant b\}$, where ω is the probability- (or sample-) space variable, and $Y(t,\omega)$, for fixed t, is a (good) function on some abstract probability space Ω with points ω. In fact we can then think of Ω as a (possibly infinite-dimensional) space with one coordinate-axis for each value of ω. Also the random variable $Y(t,.) = Y_t(.)$ say can be regarded as a vector in Ω defined by having a component $Y_t(\omega)$ along the ω-axis.

This is the exact analog of what we did for time-functions over $[a, b]$, but we emphasize again the importance of not bringing the sample functions over $[a, b]$ of the random process $Y(.)$ into the geometric picture. What we have are vectors over Ω, not over $[a, b]$.

The analogy to our earlier nonstatistical examples can perhaps be made even clearer by considering simple examples in which Ω is finite-dimensional. Thus suppose Ω has six points, $\Omega = \{1,2,3,4,5,6\}$, corresponding to a die-tossing experiment. A random variable is any real-valued function on Ω, e.g., $X(\omega) = 1$ when ω is even, and $X(\omega) = 0$ when ω is odd. We can define a probability measure on Ω, e.g., $P(\omega) = 1/6$ for all ω. Then the inner-product between two random variables (vectors) $X(.)$ and $Y(.)$ can be defined as

$$\langle X, Y\rangle = \sum_{\omega=1}^{6} X(\omega)Y(\omega)P(\omega)$$

which is the exact analog of the inner-product $\langle\cdot\rangle_2$ we defined in n-space. Note also that (by fundamental theorem of expectation),

$$\langle X, Y\rangle = \Sigma\, X(\omega)Y(\omega)P(\omega) = EXY$$

which is of course consistent with the definition we used so far. Moreover, since EXY is usually calculated via the joint density function of X and Y rather than via the usually unknown probability measure $P(\omega)$ on the usually unknown space Ω, we note that in many probabilistic calculations we rarely need to explicitly bring in a probability space. It is therefore perhaps of some interest that the notion is helpful in getting a geometric picture of the least-squares estimation problem. [Of course, as we have seen, even here the probability space need not have been introduced. But the concept cannot be avoided when one talks about, for example, the "strong" law of large numbers and how it differs from the "weak" law. But that is another story].

Exercise 2.3.1.

Given a process $y(.)$, find the 1.1.s.e. estimate of its integral $\int_0^T y(t)\, dt$ in terms of its values at the end points of the interval

When the process has covariance function $e^{-\alpha|t|}$, show that the estimate is

$$\frac{1}{\alpha}\tanh\frac{\alpha T}{2}\,(y(0)+y(T)).$$

What is the estimate for very small T ?

Exercise 2.3.2.

A discrete-time random process $\{y_i\}$ satisfies the difference equation

$$y_n + a_1 y_{n-1} + \ldots + a_k y_{n-k} = e_n$$

where the $\{e_i\}$ are uncorrelated random variables. Show that

$$\hat{y}_{n|n-1} = -(a_1 y_{n-1} + \ldots + a_k y_{n-k})$$

$$\hat{y}_{n|n-2} = -a_1 \hat{y}_{n-1|n-2} - (a_2 y_{n-2} + \ldots + a_k y_{n-k}).$$

Assume that the effect of initial conditions has died out, or equivalently that $z^n + a_1 z^{n-1} + \ldots . a_k$ is stable and that observations begin in the remote past.

Exercise 2.3.3.

Let $x(t,\omega)$ be a wide-sense stationary process with covariance function

$$R(\tau) = e^{-\alpha|\tau|}$$

Suppose that we observe this process at times T, 2T and 3T and wish to predict $x(.\,,\omega)$ at a future time, say $3T+\Delta$, $\Delta > 0$, in the least-squares sense.

Find the 1.s.e. of $x(3T+\Delta,\omega)$ in terms of $x(T,\omega)$, $x(2T,\omega)$, $x(3T,\omega)$, and the corresponding m.s.e. What happens as $\Delta\to\infty$?
NOTE: It is a striking fact that the 1.s.e. depends only on $x(3T)$. Can you show that in general the l.s.e. of $x(t+\Delta,\omega)$ in terms of $\{x(\tau,\omega),\ \sigma\leqslant\tau\leqslant t\}$ depends only on $x(t,\omega)$, the most recent observation ?

Exercise 2.3.4.

Repeat the previous problem for a process with covariance

$$R(t,s)\triangleq\min(t,s)=\begin{cases} t, & t\leqslant s\\ s, & t\geqslant s\end{cases}.$$

Exercise 2.3.5.

1. Innovations. Given a sequence of random variables $\{y_0, y_1,\ldots\}$, define

$$e_i=y_i-\hat{y}_{i|i-1},\quad \hat{y}_{0|-1}=0$$
$$\hat{y}_{i|i-1}=\text{1.1.s.e. of } y_i \text{ given } y_0,\ldots,y_{i-1}.$$

a) Show that the e_i are orthogonal random variables. (Hint: recall Gram-Schmidt)

b) Show that if $Ee_i^2>0$, all i, then the vectors $Y_n'=\lfloor y_n,\ldots,y_1,y_0\rfloor$, $E_n'=\lfloor e_n,\ldots,e_0\rfloor$ can be related, for all n, by a nonsingular triangular matrix, say $Y_n=\mathcal{S}_nE_n$.

c) If $\mathcal{H}_n$ is the linear operation that yields $\hat{Y}_n$ from Y_n, show that

$$\mathcal{H}_n=I-\mathcal{S}_n^{-1},\ \hat{Y}_n=Y_n-E_n.$$

d) Let $R_{ye}(i,j)-E\,y(i)e'(j)$ and so on. Show that

$$\mathcal{S}_n=\begin{bmatrix}1 & R_{ye}(n,n-1)R_{ee}^{-1}(n-1,n-1)\ldots & R_{ye}(n,0)R_{ee}^{-1}(0,0)\\ & 1\quad\ddots & \vdots\\ & \ddots & R_{ye}(1,0)R_{ee}^{-1}(0,0)\\ O & & 1\end{bmatrix}$$

e) Show that

$$R_{Y_n}=\mathcal{S}_nR_{E_n}\mathcal{S}_n'$$

Note that $R_{Y_n}^{-1}$ is often most easily found by inverting the above expression.

f) If **X** is a related random vector, show that

$$\hat{X}_{|n} = R_{XY_n} R_{Y_n}^{-1} Y_n = R_{XE_n} R_{E_n}^{-1} E_n$$

g) Hence, show that we can write

$$\hat{X}_{|n+1} = \hat{X}_{|n} + E\{Xe_{n+1}\} R_{ee}^{-1}(n+1, n+1) e_{n+1} .$$

This formula is the basis of the Kalman filter (cf. Section 5).

Exercise 2.3.6.

Explicitly carry out the various calculations in the previous exercise for a process $\{y_i\}$ with covariance

$$E\, y_i y_j = \rho^{|i-j|}, \quad 0 < \rho < 1 .$$

In particular, find $\mathcal{S}_n$ and $\mathcal{S}_n^{-1}$ and use this to compute $R_{Y_n}^{-1}$. Compare with a direct evaluation of $R_{Y_n}^{-1}$ and of $\mathcal{S}_n$ by inversion and triangular factorization of R_{Y_n}.

2.4. Multivariate Problems

So far the most complicated problem we have discussed involved the estimation of a single random variable X based on observations of a single random process Y(.).

It is easy to handle the case of a vector process **Y**(.) Clearly we shall have

$$\hat{X} = \int_a^b \mathbf{h}(\tau)\mathbf{Y}(\tau)\, d\tau$$

where **h**(.) obeys the vector integral equation

$$\int_a^b \mathbf{h}(\tau)\mathbf{R}_{YY}(t, \tau)\, d\tau = \mathbf{R}_{XY}(t), \quad a \leq t \leq b .$$

The question of vector **X** is just a bit more delicate. We can compute the least-squares estimate, $\hat{X}_i$, of each component of X based on observations of **Y**(.), and then we can define

$$\hat{\mathbf{X}} = [\hat{X}_1, \hat{X}_2, \ldots, \hat{X}_n] .$$

The interesting thing is that the joint statistics of $\{X_i\}$ do not enter the problem.

This happens also in another popular version of the vector problem, where we seek to estimate an **arbitrary** linear combination $\Sigma c_i X_i$ of the components of **X**. However because the $\{c_i\}$ are arbitrary, it is easy to see that

$$\widehat{(\Sigma c_i X_i)} = \Sigma c_i \hat{X}_i$$

where as above,

$$\hat{X}_i = \text{the l.s.e. of } X_i \text{ given } Y(\cdot)$$

Therefore to summarize, to find the l.s.estimate of a p-vector **X** based on observations of an m-vector Y(.) we shall have

$$\hat{X} = \int_a^b H(\tau)Y(\tau)\, d\tau$$

where H(.) is a p x m matrix satisfying the matrix integral equation

$$\int_a^b H(\tau)R_{YY}(t,\tau)\, d\tau = R_{XY}(t), \quad a \leqslant t \leqslant b$$

We could think of obtaining this integral equation by using the orthogonality condition

$$X - \hat{X} \perp Y(t), \quad a \leqslant t \leqslant b$$

with

$$\langle X, Y \rangle \triangleq EXY'$$

It is easy to check that this leads to the correct equation, but it must be noted that $\langle X, Y \rangle$ as defined above is not an inner product because for vector X **and** Y,

$$EXY' \neq EYX'$$

However, if we ignore this deficiency, the projection theorem can be used as a quick mnemonic way of obtaining the appropriate optimality conditions, and this is often done. However, in case of doubt one should always return to the arguments given at the beginning of this subsection. [There is a rigorous theory in which inner products can be matrices, see, e.g., Goldstine and L.P. Horwitz, Math. Annalen, 164 (1966).]

2.5. Smoothing for Stationary Processes

As a simple but not insubstantial example of the previous results, and also as a useful prelude to the more difficult problems to be treated in later sections, we shall study here a simple smoothing problem in which explicit solutions can be easily obtained by Fourier transformation.

Suppose $X = z(t_0)$, where z(.) is some random process whose cross-covariance with y(.) is known,

$$R_{zy}(t,\tau) = E\, z(t)y(\tau)\ .$$

Then the integral equation of Section 2.2. can be written

$$R_{zy}(t_0, t) = \int_a^b h(t_0, \tau)\ R_{yy}(\tau, t)\ d\tau$$

where we have shown explicitly the fact that the weighting h(·) will in general depend upon the point t_0 at which the estimate of z(.) is desired. However, in certain cases, $h(t_0,\tau)$ will turn out to depend only on the difference $t_0 - t$. This happens when the interval (a, b) is infinite or semi-infinite $(-\infty\ ,t_0)$ and the processes z(.) and y(.) are jointly (wide-sense) stationary, i.e.,with an abuse of notation

$$R_{yy}(\tau, t) = R_{yy}(\tau - t)$$

$$R_{zy}(\tau, t) = R_{zy}(\tau - t)\ .$$

The physical meaning of this assumption need not be explored here. The mathematical consequence will be that the integral equation will reduce to a form that can be easily solved by Fourier transformation. The equation becomes

$$R_{zy}(t_0 - t) = \int_{-\infty}^{\infty} R_{yy}(\tau - t) h(t_0, \tau)\ d\tau,\quad -\infty \leqslant t \leqslant \infty$$

We can simplify this by certain naturally suggested changes of variable. Thus let

$$t_0 - t = t'$$

Then

$$R_{zy}(t') = \int_{-\infty}^{\infty} h(t_0, \tau)\ R_{yy}(\tau + t' - t_0)\ d\tau, \ -\infty \leqslant t' \leqslant \infty$$

Now let

$$\tau + t' - t_0 = \tau'$$

Then

$$R_{zy}(t') = \int_{-\infty}^{\infty} h(t_0, t_0 + \tau' - t')\ R_{yy}(\tau')\ d\tau',\ -\infty \leqslant t' \leqslant \infty$$

Now t_0 does not appear anywhere other than in h(.,.). Therefore the value of h(. , .) must be independent of the actual value of t_0, i.e.,

$$h(t_0 + \alpha, t_0 + \alpha + \tau' - t') = h(t_0, t_0 + \tau' - t')\ \text{for all}\ \alpha .$$

In particular

$$h(t_0, t_0 + \tau' - t') = h(t' - \tau', 0),$$

so that h(.,.) is a function only of the difference of its arguments, which by abuse of notation we shall continue to denote as h(.). The integral equation now is

$$R_{zy}(t) = \int_{-\infty}^{\infty} h(t-\tau)R_{yy}(\tau)d\tau = \int_{-\infty}^{\infty} h(\tau)R_{yy}(t-\tau)d\tau .$$

The integral now has the form of a convolution integral, so that assuming that all the functions have Fourier transforms we can solve the equation as

$$h(\cdot) = \mathscr{F}^{-1}\{\mathscr{H}(f)\} = \mathscr{F}^{-1}\left\{\frac{S_{zy}(f)}{S_{yy}(f)}\right\}$$

where $\mathscr{F}\{\cdot\}$ and $\mathscr{F}^{-1}\{\ \}$ denote the Fourier transform and its inverse and

$$\mathscr{H}(f) = \mathscr{F}\{h(t)\} = \int_{-\infty}^{\infty} h(t) \exp -(i2\pi ft)\ dt$$

$$S_{zy}(f) = \mathscr{F}\{R_{zy}(t)\},\ S_{yy}(f) = \mathscr{F}\{R_{yy}(t)\}.$$

Recall also that

$$\hat{x} = \hat{z}(t_0) = \int_{-\infty}^{\infty} h(t_0 - \tau)\ y(\tau)\ d\tau$$

so that the estimate at any time t_o can be regarded as the output at time t_o of a linear time-invariant filter with impulse response h(.) and input y(.).

To get more insight into the nature of the optimum filter h(.), we shall consider some further specializations of the above problem.

Signal in Noise

Thus let

$$y(t) = z(t) + n(t)$$

where n(.) is an additive noise process, with

$$E\ n(t+\tau)n(t) = R_{nn}(\tau),\quad \mathscr{F}\{R_{nn}(t)\} = S_n(f)$$

and uncorrelated with the signal process z(.), i.e.,

$$E\ n(t)z(\tau) \equiv 0,\ \text{all } t \text{ and } \tau .$$

Then

$$\begin{aligned} R_{yy}(\tau) &= E\ y(t+\tau)y(t) \\ &= E\ z(t+\tau)z(t) + 0 + 0 + E\ n(t+\tau)n(t) \\ &= R_{zz}(\tau) + R_{nn}(\tau) \end{aligned}$$

$$S_{yy}(f) = S_{zz}(f) + S_{nn}(f)$$

$$R_{zy}(\tau) = E\ z(t+\tau)y(t) = E\ z(t+\tau)z(t) + 0 = R_{zz}(\tau)$$

$$S_{zy}(f) = S_{zz}(f)$$

and the optimum filter now is

$$\mathcal{H}(f) = \frac{S_{zz}(f)}{S_{zz}(f) + S_{nn}(f)}$$

The mean-square error with the optimum filter can be calculated as

$$
\begin{aligned}
E\ e^2_{min} = E[z(t) - \hat{z}(t)]^2 &= E[z(t) - \hat{z}(t)]z(t) - \underbrace{E[z(t) - \hat{z}(t)]\hat{z}(t)}_{=0} \\
&= R_{zz}(0) - \int_{-\infty}^{\infty} h_{opt}(\tau)E[y(t-\tau)z(t)]\ d\tau \\
&= R_{zz}(0) - \int_{-\infty}^{\infty} h_{opt}(\tau)R_{zy}(\tau)\ d\tau
\end{aligned}
$$

The first term

$$R_{zz}(0) = E\ z^2(t) = \text{variance of } z(\cdot)$$

represents the error that would be incurred if no observations were made, in which case

$$\hat{z}(t) = E\ z(t) = 0, \quad E[z(t)]^2 = E\ z^2(t)$$

The second term

$$\int h(\tau)R_{zy}(\tau)\ d\tau$$

is the reduction in the mean-square error that can be obtained by using the optimum linear-least squares filter. In many applications, the important thing is not so much to exactly implement the optimum filter as it is to determine the magnitude of the maximum possible reduction in mean-square error. If this is not very large, there may be no point in using the optimum filter. One can also compare the reduction provided by simple nonoptimum filters with that provided by the optimum filter. For such purposes, it is useful to have an expression for the minimum mean-square error that does not require knowledge of the optimum filter. Such an expression can be derived in the frequency domain.

By use of Parseval's theorem, we can write the error as

$$E\, e^2_{min} = \int_{-\infty}^{\infty} S_z(f)\, df - \int_{-\infty}^{\infty} \mathcal{H}_{opt}(f) S_{zy}(-f)\, df$$

In the previously studied special case where

$$y(t) = z(t) + n(t)\,, \quad E\; z(t)n(\tau) \equiv 0$$

this reduces to the nice form

$$E\, e^2_{min} = \int_{-\infty}^{\infty} \frac{S_{zz}(f) S_{nn}(f)}{S_{zz}(f) + S_{nn}(f)}\, df$$

Example 2.5.1

Suppose that we have uncorrelated signal and noise processes with

$$S_n(f) = N_0/2 \quad, \quad -\infty < f < \infty$$

$$S_z(f) = \begin{cases} A(1 - |f|/f_c), & |f| < f_c \\ 0 & , \text{elsewhere} \end{cases}.$$

i) Calculate and sketch the transfer function of the optimum smoothing filter.

ii) Calculate the minimum mean-square-error and also its limiting values for low signal to noise ratio (SNR) – where $N_0 \longrightarrow \infty$ while A is fixed, and high SNR where $A \longrightarrow \infty$ for fixed N_0.

iii) Compare the results in (i) and (ii) with those for the ideal low pass filter,

$$I(f) = \begin{cases} 1\,, & |f| < f_c \\ 0, & \text{elsewhere} \end{cases}.$$

The filter I(f) passes the signal process "undistorted", while rejecting all noise outside the (frequency) range of the signal. This was the "classical" solution before the advent of the statistical theory.

Solution

i) We have

$$H_{opt}(f) = \begin{cases} 0 & |f| \geqslant f_c \\ \dfrac{2A}{N_0} \dfrac{[1 - |f|/f_c]}{1 + \dfrac{2A}{N_0}[1 - \dfrac{|f|}{f_c}]}\,, & |f| < f_c \end{cases}.$$

The optimum filter rejects all inputs outside the frequency range of the signal process. However, unlike the ideal low pass filter I(f), we do not weigh all frequencies within the signal range equally, a requirement that matches the criterion of leaving the signal process "undistorted". On the contrary, we have a differential weighting, roughly proportional to the expected (i.e., average) signal to noise ratio $(S_z(f)/N_0)$ at each frequency. This is very clearly seen at low signal-to-noise ratios,

$$H_{opt}(f) \to \frac{2A}{N_0}[1 - |f|/f_c], \quad |f| < f_c$$
$$\text{as } A/N_0 \to 0 .$$

On the other hand when $A/N \longrightarrow \infty$, we return to the "distortionless" filter

$$H_{opt}(f) \to 1, \quad |f| < f_c .$$

For intermediate values of A/N_0 we have a trade-off between the weightings in these two limiting cases.

ii) We can readily calculate

$$E\, e^2_{min} = N_0 f_c \left[1 - \frac{N_0}{2A} \ln \left(1 + \frac{2A}{N_0}\right)\right]$$

Now, for low SNR (A fixed, $N_0 \longrightarrow \infty$) we see that (some care is needed – we must keep at least two terms in the expansion of $\ln(1 + 2A/N_0)$)

$$E\, e^2_{min} \to Af_c ,$$

while for high SNR ($A \longrightarrow \infty$, N_0 fixed),

$$E\, e^2_{min} \to N_0 f_c .$$

These formulas have simple interpretations. Af_c turns out to be the variance of the signal process z ($R_z(0) = \int S_z(f)\, df$), which is what we would expect to get since $H_{opt}(f) \longrightarrow 0$ as $N_0 \longrightarrow \infty$ and therefore $\hat{z} \longrightarrow 0$. On the other hand as $A \longrightarrow \infty$ for fixed N_0, $H_{opt}(f) \longrightarrow I(f)$ and the mean-square-error is just equal to the noise power in the "passband" $(-f_c, f_c)$.

Incidentally, we may remark that $N_0 f_c$ is the mean-square-error achievable with the ideal low pass filter $I(f)$ for all values of A and A/N_0. (Prove this). From plots as shown on page 37, we can evaluate the potential advantage of the optimum filter at low signal to noise ratios (which is essentially when one would want to use an optimum solution anyway).

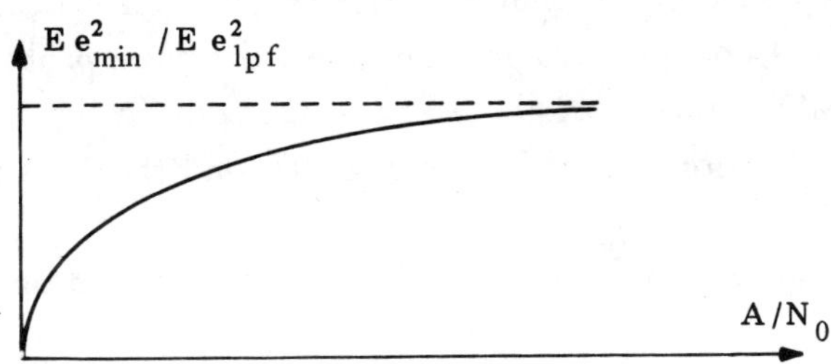

Exercise 2.5.1

Let

$$R_{zz}(t,\tau) = A\, e^{-\alpha|t-\tau|} \cos \omega_0(t-\tau)$$

$$R_{nn}(t,\tau) = (N_0/2)\delta(t-\tau)$$

Then if $\alpha \ll \omega_0$, show that we shall have

$$S_{zz}(f) = \frac{A\,\alpha/2\pi}{\alpha^2 + (\omega-\omega_0)^2}\,, \quad \omega = 2\pi f$$

$$S_{nn}(f) = N_0/2$$

$$\mathcal{H}_{opt}(f) = \frac{A\alpha/\pi}{2\,A\alpha + N_0\alpha^2 + N_0(\omega - \omega_0)^2}$$

$$E\,e^2_{min} = \frac{A}{\sqrt{1+\frac{2A\cdot 2\pi}{N_0\alpha}}}$$

3. WIENER FILTERS

Given two zero-mean jointly stationary random processes x(.) and y(.) with known auto- and cross- covariance functions, the basic problem studied by N. Wiener [1] at the MIT Radiation Laboratory in the early forties, was the following:

Given observations $\{y(\tau), -\infty < \tau < t\}$ find the 1.1.s.e. of $x(t+\lambda)$, λ a fixed constant.

If we denote the estimate by $\hat{x}(t+\lambda|t)$, the task is to find $h(t,\tau)$ such that

$$\hat{x}(t+\lambda) = \int_{-\infty}^{t} h(t,\tau)y(\tau)\, d\tau$$

and

$$E[x(t+\lambda) - \hat{x}(t+\lambda)]^2 = \text{minimum}, \quad -\infty < t < \infty$$

The function $h(t,\tau)$ can be regarded as the impulse response of a causal (i.e., $h(t,\tau) = 0$, $t < \tau$), linear system whose response at time t to the input waveform $\{y(\tau), -\infty < \tau < t\}$ is the desired estimate $\hat{x}(t+\lambda|t)$. It will be shown presently that the assumption of jointly stationary x(.) and y(.) shows that the optimum system is in fact time-invariant, i.e., $h(t,\tau)$ is of the form $h(t-\tau)$.

3.1. The Wiener-Hopf Equation

To determine the optimum system, we use the orthogonality property (projection theorem for linear least-squares estimates),

$$x(t+\lambda) - \hat{x}(t+\lambda|t) \perp y(\sigma), \quad -\infty < \sigma < t$$

which yields

$$E[x(t+\lambda)y(\sigma)] = \int_{-\infty}^{t} h(t,\tau)E[y(\tau)y(\sigma)]\, d\tau$$

or

$$R_{xy}(t+\lambda-\sigma) = \int_{-\infty}^{t} h(t,\tau)R_y(\tau-\sigma)\, d\tau, \qquad -\infty < \sigma < t$$

$$= \int_0^{\infty} h(t, t-\tau)R_y(t-\sigma-\tau)\, d\tau, \quad -\infty < \sigma < t.$$

If we change the variable $t-\sigma$ to t the equation becomes

$$R_{xy}(t+\lambda) = \int_0^{\infty} h(t+\sigma, t+\sigma-\tau)R_y(t-\tau)\, d\tau, \qquad t > 0.$$

The left hand side of this equation does not depend upon the value of σ, nor does $R_y(t-\tau)$. Therefore $h(t+\sigma, t+\sigma-\tau)$ must be a function only of the difference of its arguments, which by an abuse of notation we shall still write as h(.),

$$h(t+\sigma, t+\sigma-\tau) = h(\tau),$$

The equation for h(.) is

$$R_{xy}(t+\lambda) = \int_0^\infty h(\tau)R_y(t-\tau)\, d\tau, \quad t>0.$$

Since

$$h(\tau) = 0, \quad \tau < 0$$

we can also write

$$R_{xy}(t+\lambda) = \int_{-\infty}^{\infty} h(\tau)R_y(t-\tau)\, d\tau, \quad t>0 \tag{1}$$

In this form it might appear that the equation can be readily solved by taking Fourier transforms, but the presence of the constraint $t > 0$ prevents this. The easiest way of seeing why taking Fourier or even bilateral Laplace Transforms would not work is to try it. Thus if

$$H(s) = \int_{-\infty}^{\infty} h(t)e^{-st}\, dt$$

$$S_y(s) = \int_{-\infty}^{\infty} R_y(t)e^{-st}\, dt,$$

then we can write

$$\int_0^\infty R_{xy}(t+\lambda)e^{-st}\, dt = \int_{-\infty}^{\infty} d\tau\, h(\tau)e^{-s\tau} \int_0^\infty R_y(t-\tau)e^{-s(t-\tau)}\, dt$$

$$= \int_{-\infty}^{\infty} d\tau\, h(\tau)e^{-s\tau} \int_{-\tau}^{\infty} R_y(\sigma)e^{-s\sigma}\, d\sigma$$

The presence of the $-\tau$ in the limits of the integral over t prevents any easy solution. The $-\tau$ appears because the range of integration is restricted by the constraint $t > 0$; if there were no constraints, there would be no difficulty. (As we shall see later, there will also be no difficulty if y(.) is white, so that $R_y(\sigma) = \delta(\sigma)$) .

Therefore a more sophisticated technique is needed to solve equation (1) and such a technique was developed by Wiener and Hopf in 1931; in fact, though special cases of equations of the form (1) had been encountered earlier (by Hvolson in 1894 in a problem of light scattering and by Milne in 1921 in a problem of radiative transfer), since 1931 such equations have been known as Wiener-Hopf equations. We shall present here(*) a somewhat special form of the Wiener-Hopf

(*) We should mention that there are other methods of solving Wiener-Hopf (and related) equations — see Sections 7, 8, 9.

method as applied to stationary processes with rational power spectral density. Such processes are also known as lumped processes, and for a variety of reasons, it will be desirable to examine some of their properties before proceeding with the solution of the Wiener-Hopf equation.

3.2. Lumped Processes

A lumped process will be one that can be obtained by passing white noise through a lumped or finite-dimensional linear system.

If the linear system is stable and time-invariant with transfer function $H(s)$, if the input white noise has unit intensity, and if all transients have died out, then the bilateral Laplace transform of the covariance function of the system output, say $y(.)$, is

$$S_y(s) = H(s)H(-s)$$

Since the system is lumped we know that

$$H(s) = \text{ a ratio of polynomials in } s$$

and that

$$S_y(s) = \text{ a ratio of polynomials in } s^2$$

The inverse Laplace Transform of $S_y(s)$ is the covariance function and this clearly has the form

$$R_y(t) = \sum_{i=1}^{n} c_i e^{-d_i|t|}$$

where the c_i are real and corresponding to any complex d_i there also occurs its conjugate d_i^* (purely imaginary d_i are excluded). For the present, we shall work mostly with $S_y(s)$, but shall return to $R_y(t)$ when we study extensions of the Wiener filtering problem.

The fact that $R_y(t)$ is a sum of **damped** exponentials for positive as well as negative t means that the region of definition of its bilateral Laplace transform, $S_y(s)$, includes the $j\omega$-axis. In fact, when $s = j\omega$, $S_y(s)$ reduces to

$$S_y(j\omega) \triangleq S_y(\omega) = \text{the power spectral density function in the usual sense.}$$

However for convenience we shall also sometimes refer to $S_y(s)$ as the power-spectral density function of $y(.)$, though this of course is an abuse (and not only of notation): $S_y(s)$ will not in general be nonnegative or even real, but it must obey the condition

$$S_y(s) = S_y(-s).$$

On the other hand, we must have (for a real process $y(.)$) that

$$S_y(\omega) \text{ is real, even in } \omega \text{, and nonnegative.}$$

This fact means that the poles and zeros of $S_y(s)$ (i.e., roots of the denominator and numerator of $S_y(s)$) have a particular quadrantal symmetry when plotted in the complex plane. We observe that the pole-zero configuration

i) is symmetric about the σ-axis, because $S_y(\omega)$ is real,

ii) is symmetric about the $j\omega$-axis, because $S_y(\omega)$ is even,

iii) has $j\omega$-axis zeros of even multiplicity, because $S_y(\omega) \geqslant 0$,

iv) has no $j\omega$-axis poles, because then the inverse Fourier transform cannot be a covariance function. [Check this for the function $1/\omega^2$.]

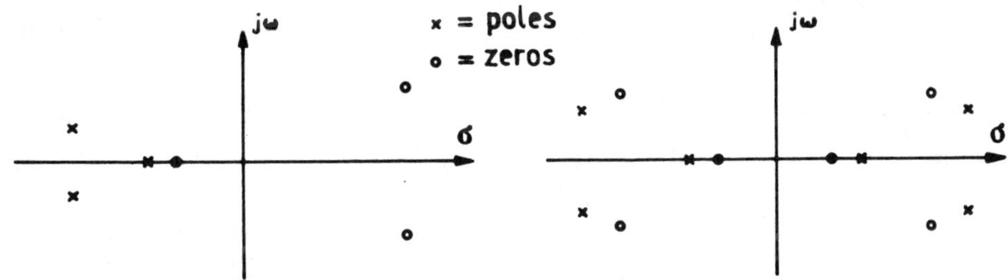

Fig. 1: Pole-zero plot of H(s) — note that there are no $j\omega$-axis poles because of the assumption of stability. The absence of right-half-plane poles is due to the (nonessential) assumption of a causal system (recall that the contour of integration is being taken, by hypothesis, to be the $j\omega$-axis.

Fig. 2: Pole-zero plot of $S_y(s)$. All properties follow from the fact that $S_y(s) = H(s)H(-s)$.

Canonical Factorization

For convenience in further discussions we shall assume that the numerator and denominator polynomials of $S_y(s)$ have leading coefficient equal to unity. Then we shall denote

$$\{z_i\}_1^m = \text{left-half-plane (LHP) zeros of } S_y(s)$$

$$\{p_i\}_1^n = \text{LHP poles of } S_y(s), \quad n \geqslant m\,.$$

Given a particular $S_y(s)$ we can factor it in several ways into the form

$$S_y(s) = H(s)H(-s)$$

where

H(s) = the transfer function of a stable and causal system.

The constraints on H(s) require that all poles of H(s) be in the LHP, but they do not constrain the zeros. A factorization of the form

$$S_y(s) = S_y^+(s)S_y^-(s)$$

where

$$S_y^+(s) = \prod_1^m(s - z_i)/\prod_1^n(s - p_i)$$

and

$$S_y^-(s) = S_y^+(-s)$$

will be said to be **canonical.** $S_y^+(s)$ will be called the **canonical spectral factor** of $S_y(s)$. The reasons for the name will become clear as we proceed but we may note here that the most important property of the canonical factor is that not only $S_y^+(s)$ but both $S_y^+(s)$ and its **inverse** $\frac{1}{S_y^+(s)}$ are transfer functions of lumped **causal** systems.

Note that we could always throw into $S_y^+(s)$ a multiplier of the form $e^{s\tau}$, τ fixed, without affecting $S_y(s)$, but this would mean that $S_y^+(s)$ would no longer be the transfer function of a lumped system.

We remark that in network theory, a causal transfer function such as $S_y^+(s)$ that has all its zeros in the LHP is said to be a **minimum-phase** function; the point is that if we used a RHP zero in place of LHP zero, the magnitude of the transfer function for $s = j\omega$ would not be changed but its "phase angle" for $s = j\omega$ (for $\omega \geqslant 0$) would be increased.

Remark: For those familiar with the network theoretical results on the Hilbert transform relations between the "gain" and "phase" of causal minimum phase systems, we mention that the canonical factorization can also be carried out for nonrational power spectral densities obeying the so-called Paley-Wiener condition that

$$\int_{-\infty}^{\infty} \frac{|\ln S_y(\omega)|}{1+\omega^2} d\omega < \infty$$

In this case, the canonical causal factor can be found as

$$S_y^+(\omega) = \sqrt{S_y(\omega)}\, e^{j\Theta(\omega)}$$

where $\Theta(\omega)$ is the Hilbert transform of $\ln\sqrt{S_y(\omega)}$.

Additive Decompositions

We shall need one more notational convention.

Let f(t) be a not necessarily symmetric time function whose (bilateral)

Laplace transform exists in a region containing the $j\omega$-axis. The covariance functions and cross-covariance functions of lumped processes and transfer functions of lumped systems all satisfy this constraint.

If

$$\mathcal{L}[f(t)] = \int_{-\infty}^{\infty} f(t)e^{-st}\,dt = F(s), \quad \text{say}$$

then we shall write

$$\mathcal{L}[f(t)1(t)] = \{F(s)\}_+ \tag{2a}$$

$$\mathcal{L}[f(t)1(-t)] = \{F(s)\}_- \tag{2b}$$

where

$$1(t) = \begin{cases} 1, & t \geqslant 0 \\ 0, & t < 0 \end{cases}$$

When F(s) is rational, it can be shown that

$$\mathcal{L}[f(t)1(t)] = \sum_{\substack{\text{LHP} \\ \text{Poles}}}^{*} \{\text{partial fraction expansion of } F(s)\}$$

$$\mathcal{L}[f(t)1(-t)] = \sum_{\substack{\text{RHP} \\ \text{Poles}}} \{\text{partial fraction expansion of } F(s)\}$$

By $\sum^{*}$ we mean the sum is to include any constants or any finite sums of positive powers of s that may be present in F(s) if F(s) is not a proper fraction (cf. Example 3 below). If F(s) is not rational then $\{F(s)\}_+$ must be computed from (2) (see Example 2 below) or, equivalently, from the formula (3) below.

A method of establishing the relations for $\{F(s)\}_\pm$ is to consider the integral formula for computing the inverse Laplace transform

$$f(t) = \frac{1}{2\pi j}\int_{-j\infty}^{j\infty} F(s)e^{st}\,ds\,.$$

This integral can be evaluated (under certain conditions that include the case of rational F(s)) by Cauchy's residue theorem and Jordan's lemma by closing the contour of integration along a large circular arc in the RHP for $t < 0$ and in the LHP for $t > 0$. Clearly f(t) will be zero for $t < 0$ ($t > 0$) if F(s) has no poles in the RHP (LHP). Note that whether F(s) is rational or not, we can always write

(3) $$\{F(s)\}_+ = \int_{0-}^{\infty} dt\, e^{-st} \int_{-j\infty}^{j\infty} F(s)e^{st} \frac{ds}{2\pi j}$$

Example 3.2.1: If

$$R(t) = e^{-\alpha|t|}, \quad \alpha > 0$$

then

$$R(t)1(t) = e^{-\alpha|t|}1(t)$$

and

$$\mathcal{L}[R(t)1(t)] = \frac{1}{s+\alpha}, \quad \text{Re } s > -\alpha$$

Now

$$S_y(s) = \frac{2\alpha}{\alpha^2 - s^2} = \frac{1}{s+\alpha} - \frac{1}{s-\alpha}, \quad -\alpha < \text{Re } s < \alpha$$

so that

$$\{S_y(s)\}_+ = \frac{1}{s+\alpha} = \mathcal{L}[R(t)1(t)] .$$

Example 3.2.2 : If

$$f(t) = \begin{cases} \exp -\alpha(t+\lambda), & t > -\lambda, \quad \alpha > 0 \\ 0, & t < -\lambda \end{cases}$$
$$= \exp -\alpha(t+\lambda)\cdot 1(t+\lambda)$$

then its transform is not a rational function of s. We cannot now use partial fractions directly, but can proceed as follows. We note that

$$F(s) = e^{s\lambda}/(s+\alpha), \quad \text{Re } s > -\alpha$$

and

$$f(t)1(t) = e^{-\alpha\lambda}e^{-\alpha t}1(t),$$

so that

$$\{F(s)\}_+ = \frac{e^{-\alpha\lambda}}{s+\alpha}, \quad \text{Re } s > -\alpha$$

This special formula is worth remembering.

Example 3.2.3 : Let

$$f(t) = \delta(t) + (\alpha - \beta)e^{-\beta t}1(t), \quad \beta > 0$$

Then

$$F(s) = \frac{s+\alpha}{s+\beta} = \{F(s)\}_+ .$$

3.3. The Wiener-Hopf Technique

The Wiener-Hopf equation is

$$R_{xy}(t+\lambda) = \int_0^\infty h(\tau)R_y(t-\tau)\, d\tau, \quad t > 0$$

where

$$h(t) = 0, \quad t < 0$$

We know (cf. the discussions at the end of Sec. 3.2) that this means that H(s), the (bilateral) Laplace transform of h(t), will have no poles in the RHP.

Let the transform of $R_{xy}(t)$ be

$$S_{xy}(s) = \int_{-\infty}^{\infty} R_{xy}(t)e^{-st}\, dt$$

and note that

$$S_{xy}(s)e^{s\lambda} = \int_{-\infty}^{\infty} R_{xy}(t+\lambda)e^{-st}\, dt$$

We shall prove that, in the notation of Sec. 2.2.,

$$H(s) = \frac{1}{S_y^+(s)} \left\{ \frac{S_{xy}(s)e^{s\lambda}}{S_y^-(s)} \right\}_+ \tag{4a}$$

$$= \frac{1}{S_y^+(s)} \int_0^\infty dt\, e^{-st} \int_{-j\infty}^{j\infty} \frac{S_{xy}(s)e^{s\lambda}}{S_y^-(s)} e^{st} \frac{ds}{2\pi_j} \tag{4b}$$

Proof: Let

$$g(t) = R_{xy}(t+\lambda) - \int_0^\infty h(\tau)R_y(t-\tau)\, d\tau$$

Then, from (1), we have

$$g(t) = 0 \quad , \quad t \geq 0$$
$$= \text{unknown} \; , \quad t < 0 .$$

However, we know (cf. the discussion at the end of Sec. 3.2) that for such a function

$$G(s) = \mathcal{L}[\, g(t)\,] = \text{ a function with no poles in the LHP}$$

This simple property of Laplace transforms, combined with the canonical factorization of $S_y(s)$, is the key to the Wiener-Hopf method.

For we observe that

$$G(s) = S_{xy}(s)e^{s\lambda} - H(s)S_y(s)$$

and, using the canonical factorization of $S_y(s)$, we can write

$$(5) \qquad \underbrace{\frac{G(s)}{S_y^-(s)}}_{\substack{\text{no poles}\\ \text{in LHP}}} = \underbrace{\frac{S_{xy}(s)e^{s\lambda}}{S_y^-(s)}}_{\substack{\text{poles in RHP}\\ \text{and LHP}}} - \underbrace{H(s)\,S_y^+(s)}_{\substack{\text{no poles in}\\ \text{RHP}}}.$$

We note that $G(s)/S_y^-(s)$ has no LHP poles because $G(s)$ has none and $S_y^-(s)$ has no LHP zeros; similarly because $h(t) = 0$, $t < 0$, $H(s)$ has no RHP poles and neither has $S_y^+(s)$ or $H(s)\,S_y^+(s)$. The above equation then shows that for balance (or by applying the operation $\{\cdot\}_+$ to both sides) we must have

$$H(s)S_y^+(s) = \left\{ \frac{S_{xy}(s)e^{s\lambda}}{S_y^-(s)} \right\}_+$$

which gives the stated formula (4) for H(s).

Example 3.3.1 : Pure Prediction

If $x(t) \equiv y(t)$ and $\lambda > 0$, then we have a problem of pure prediction. In this case, we can see that

$$(6) \qquad H(s) = \frac{1}{S_y^+(s)} \left\{ S_y^+(s)\, e^{s\lambda} \right\}_+$$

This formula can be simplified even further for lumped processes by appropriate use of the result of Example 3.2.2.

Example 3.3.2 : Pure Prediction of Ornstein-Uhlenbeck Process.

Let

$$x(t) \equiv y(t), \quad \lambda > 0,$$

and

$$R_{xy}(t) = R_y(t) = e^{-\alpha|t|}$$

Then

$$S_y(s) = \frac{2\alpha}{\alpha^2 - s^2}, \quad S_y^+(s) = \frac{\sqrt{2\alpha}}{\alpha + s}$$

and

$$\left\{ \frac{\sqrt{2\alpha}}{\alpha + s} e^{s\lambda} \right\}_+ = \frac{\sqrt{2\alpha}\, e^{-\alpha\lambda}}{\alpha + s}$$

Therefore

$$H(s) = e^{-\alpha\lambda}, \ h(t) = e^{-\alpha\lambda}\delta(t)$$

and

$$\begin{aligned} \hat{x}(t+\lambda) &= \int_{-\infty}^{t} e^{-\alpha\lambda}\delta(t-\tau)x(\tau)\, d\tau \\ &= e^{-\alpha\lambda}x(t) . \end{aligned} \tag{7}$$

This is such a simple result that we might expect that it could be derived more directly. This is true (see Example 2.2.1 and also Exercise 3.4.2) and in fact it was the search for such methods that slowly led to the Kalman-filter theory to be discussed later.

Example 3.3.3 : Filtering of Ornstein-Uhlenbeck Process in White Noise.

Let

$$y(t) = \sqrt{P}\, x(t) + v(t)$$

where

$$S_v(f) = N_0/2$$

$$S_x(s) = 2\alpha/(\alpha^2 - s^2)$$

and x(.) and v(.) are uncorrelated. Also let

$$\Lambda = 4P/\alpha N_0 = 2(\text{the “signal-to-noise” ratio})$$

Then some algebra shows that

$$S_y^+(s) = \sqrt{\frac{N_0}{2}} \left\{ \frac{s + \alpha\sqrt{1+\Lambda}}{s+\alpha} \right\}$$

$$S_y^-(s) = \sqrt{\frac{N_0}{2}} \left\{ \frac{\alpha\sqrt{1+\Lambda} - s}{\alpha - s} \right\}$$

and

$$S_{xy}(s) = \sqrt{P} S_x(s) = \frac{2\alpha\sqrt{P}}{\alpha^2 - s^2}$$

Thus

$$\left\{ \frac{S_{xy}(s)}{S_y^-(s)} \right\}_+ = \left\{ \sqrt{\frac{N_0}{2}} \cdot \frac{\alpha^2 \Lambda}{(\alpha + s)(\alpha\sqrt{1+\Lambda} - s)} \right\}_+$$

$$= \sqrt{\frac{N_0}{2}} \cdot \frac{\alpha\Lambda}{1 + \sqrt{1+\Lambda}} \cdot \frac{1}{(s+\alpha)}$$

Consequently

$$H(s) = \frac{1}{S_y^+(s)} \left\{ \frac{S_{xy}(s)}{S_y^-(s)} \right\}_+ = \frac{\alpha[\sqrt{1+\Lambda} - 1]}{(s + \alpha\sqrt{1+\Lambda})}$$

which in the special case $N_0/2 = 1$, $P = 1$ agrees with the Yovits and Jackson result discussed in the next example.

Example 3.3.4 : Filtering in White Noise

Let

$$y(t) = z(t) + v(t)$$

where

$S_v(f) = 1$

$S_z(f) =$ a rational function in f^2 with numerator of **lower** degree than the denominator,

and

$$E\ z(t)v(s) \equiv 0.$$

Then the optimum filter for $\lambda = 0$ turns out to be given by

$$H(s) = 1 - \frac{1}{S_y^+(s)} \tag{8}$$

and the minimum mean-square-error by

$$\text{m.s.e.} = \int_{-\infty}^{\infty} \ln(1 + S_z(f))\,df, \quad s = \sigma + j2\pi f. \tag{9}$$

These formulas were first obtained by Yovits and Jackson [2]. The formula (8) for H(s) is easy to derive and will be left to the reader – however, we may mention that an important generalization to nonstationary results will be proved later (Sec. 6); compare also with the result of Exercise 2.3.5(c). The formula (9) for the m.s.e. is more difficult and will be established in Sec. 7.

Remark: In the above we have given the Wiener-Hopf technique in its simplest form; more elaborate studies can be found in many places, but we have found Levinson [3] and Krein [4] to be especially instructive.

Exercise 3.3.1.

We have a random process Y(.) with covariance function

$$R(t) = \frac{3}{2} e^{-|t|} + \frac{11}{3} e^{-3|t|}.$$

a) Show that

$$S_y(s) = \frac{49 - 25s^2}{(1 - s^2)(9 - s^2)}$$

and that the optimum filter for predicting $\ell n\, 2$ seconds ahead has impulse response

$$h(t) = \frac{1}{5}\delta(t) + \frac{3}{25} e^{-7t/5} 1(t)$$

b) Make up an RC circuit that will have this impulse response.

c) Show that the mean-square-error when using the optimum filter is

$$\text{m.s.e.} = \frac{31}{6} - \frac{e^{-2\lambda}}{2} - \frac{8e^{-4\lambda}}{4} - \frac{16e^{-6\lambda}}{6}, \quad \lambda = \ell n\, 2$$

What would be the best prediction and corresponding m.s.e. if we had no observations ?

Exercise 3.3.2.

If $$S_x(s) = 1/(\alpha^2 - s^2)^2,$$

find λ_{max} so that the mean-square-error in predicting λ_{max} seconds ahead is not more than 10% of the variance of x(.).

Exercise 3.3.3

Establish the formula (8).

Exercise 3.3.4

Verify that (7) is correct by checking that

$$x(t+\lambda) - e^{-\alpha\lambda}\, x(t) \perp x(s), \qquad s \leqslant t\,.$$

Exercise 3.3.5.

Develop the Wiener-Hopf technique for discrete-time random sequences using z-transforms instead of Laplace transforms.

3.4. The Innovations Approach

Given some knowledge of complex variables and bilateral Laplace transforms, the Wiener-Hopf technique (for lumped processes) is quite simple. However, in the late forties such a background was not very widespread. Partly motivated by this, in 1950, Bode and Shannon [5] and Zadeh and Ragazzini [6] published descriptions of an alternative(*) approach based on certain linear system ideas.

The method of Bode-Shannon and Zadeh-Ragazzini was based on two observations:

i) the Wiener-Hopf equation is trivial to solve if the observed process is white noise,

ii) for stationary lumped processes it is easy to transform, without loss of information, an arbitrary observed process into a white-noise process.

Let us consider these points separately.

i) **Estimation Based on White-Noise Observations**

Suppose that we are to estimate $x(t+\lambda)$, given observations of a white noise process $\{\upsilon(\tau), -\infty < \tau < t\}$ with

$$R_\upsilon(t) = \delta(t)\,, \; S_\upsilon(f) = 1$$

The corresponding Wiener-Hopf equation for the optimum estimating filter is

$$R_{x\upsilon}(t+\lambda) = \int_0^\infty g(\tau) R_\upsilon(t-\tau)\, d\tau, \qquad t \geqslant 0 \tag{10}$$

But since $\upsilon(.)$ is white,

$$\int_0^\infty g(\tau) R_\upsilon(t-\tau)\, d\tau = \int_0^\infty g(\tau)\delta(t-\tau)\, d\tau = g(t)$$

so that we have, from the equation (1), that the optimum filter for estimating

(*) Actually, essentially the same approach had been developed in a different context by Wold (1938), Kolmogorov (1941), Karhunen (1947), Hanner (1950), see the discussion in [8].

$x(t+\lambda)$ from $\{v(s), -\infty < s < t\}$ is

$$g(t) = R_{xv}(t+\lambda), \quad t \geqslant 0. \tag{11}$$

Thus the optimum (causal) filter is immediately determined. Note that its Laplace transform is

$$G(s) = \{S_{xv}(s)e^{s\lambda}\}_+ \tag{12}$$

where $S_{xv}(s)$ is the Laplace transform of $R_{xv}(t)$.

ii)Conversion to White Noise

A process with power-spectral density $S_y(s)$ can be converted to a white-noise process by passing it through a filter with transfer function W(s) such that

$$W(s)W(-s) = 1/S_y(s)$$

For, if a process y(.) is passed through such a filter, the power-spectral density of the output will be

$$S_y(s).W(s)W(-s) = 1.$$

Clearly there are many filters W(s) that will "whiten" a given process. However, one requirement on W(s) in our problem is that it be causal, for otherwise an estimate based on observations of the white noise output upto time t would actually depend on observations of the input process y(.) for times greater than t. The requirement of causality on W(s) can be met by requiring that all poles of W(s) be in the left half plane. But this still does not specify W(s) uniquely.

However, note that in addition to causality we should also require that the transformation to white noise be done without "loss of information". There are various definitions of information but the property of interest here is that all linear combinations of the input random variables upto time t be in a one-to-one relationship with some linear combination of output random variables upto time t. This feature will be ensured if and only if in addition to causality of W(s) we also have causality of the inverse filter 1/W(s) but this means that both the poles and zeros must be in the LHP.

We may now recall that this requirement can be met (cf. Sec. 3.2.) if and only if we choose

$$W(s) = \frac{1}{S_y^+(s)} \tag{13}$$

where

$$S_y^+(s) = \text{the canonical spectral factor of } S_y(s).$$

We shall denote the response to y(.) of the canonical whitening filter $1/S_y^+(s)$ by

υ (.), which will be called the **innovations** (**or new information**) process of y(.). The name, apparently first used by Wiener, Masani and Kallianpur in the mid-fifties, arises from the fact that the values of υ (.) at different time instants are uncorrelated unlike the values of y(.). Therefore, the observation y(t) of y(.) at time t does not bring us completely "new" information because some information about y(t) can be obtained from other values of y(.); on the other hand the value $\upsilon(t)$ cannot be predicted from other values of $\upsilon(.)$ (or more precisely the predicted value of $\upsilon(t)$ given for example $\upsilon(s)$ is zero because $E\,\upsilon(t)\,\upsilon(s) = 0$). Therefore, each observation υ (t) brings us "new" information, and υ (.) may be called the new information or innovations process of y(.). We shall see that the determination of y(.) in terms of υ (.) and of the innovations representation of y(.) in terms of υ (.) (see Fig. 3), play a fundamental role in many estimation and detection problems involving the process y(.).

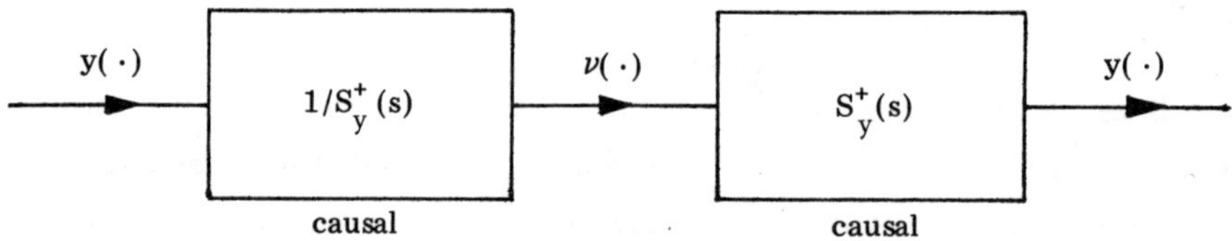

Fig. 3: The canonical whitening filter and its inverse,

the innovations representation (IR) of y(·) in terms of the innovations ν(·)

Solution of the Original Problem

We can now combine the results in i) and ii) to obtain a solution of the original problem of determining $\hat{x}(t+\lambda)$ given $\{y(\tau), -\infty < \tau < t\}$.

The optimum filter can be regarded (see Fig. 4) as a cascade of the canonical whitening filter and the optimum filter for white-noise observations:

$$H(s) = \frac{1}{S_y^+(s)} \cdot \left\{ S_{x\upsilon}(s)\, e^{s\lambda} \right\}_+ \tag{14}$$

Now by linear system relationships we know that since $\upsilon(.)$ is obtained by passing y(.) through the linear filter $1/S_y^+(s)$,

$$\begin{aligned} S_{x\upsilon} &= S_{xy}(s)/S_y^+(-s) \\ &= S_{xy}(s)/S_y^-(s)\,. \end{aligned}$$

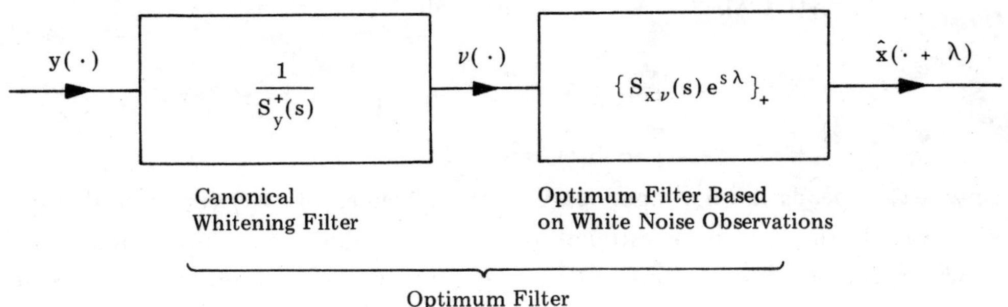

Fig. 4: The optimum filter by the innovations method

Therefore,

$$H(s) = \frac{1}{S_y^+(s)} \cdot \left\{ \frac{S_{xy}(s)\, e^{s\lambda}}{S_y^-(s)} \right\}_+$$

which is the same as the formula (4) we obtained by the Wiener-Hopf method in Sec. 3.3

An Alternative More Time-Domain Approach

The innovations method of the previous section can also be described in more purely time-domain terms. Such analyses were given by Karhunen (1949) and Hanner (1950) at the same time as the work of Bode-Shannon (1950) and Zadeh-Ragazzini (1950). These time-domain analyses have the advantage that they can be extended to nonstationary processes.

We shall rework the problem of pure prediction in which we wish to find the 1.1.s.e. $\hat{y}(t+\lambda|t)$ of $y(t+\lambda)$, $\lambda > 0$, given $\{y(\tau), -\infty \leq \tau \leq t\}$.

The first step will be to find the innovations representation of y(.). The discussion of Section 3.4 shows that this is

$$y(t) = \int_{-\infty}^{t} k(t-\tau)\upsilon(\tau)\, d\tau\,, \quad E\,\upsilon(\tau)\upsilon(\sigma) = \delta(\tau - \sigma) \tag{15}$$

where k(.) is the impulse response corresponding to the transfer function $S_y^+(s)$,

$$k(t) = \int_{-j\infty}^{j\infty} S_y^+(s)\, e^{st}\, ds/2\pi j\,. \tag{16}$$

Now the random variable $y(t + \lambda)$ that we wish to estimate can be written

$$y(t+\lambda) = \int_{-\infty}^{t+\lambda} k(t+\lambda-\tau)\nu(\tau)\, d\tau$$

$$= \int_{-\infty}^{t} k(t+\lambda-\tau)\nu(\tau)\, d\tau + \int_{t}^{t+\lambda} k(t+\lambda-\tau)\nu(\tau)\, d\tau$$

$$= a(t) + b(t), \text{ say.}$$

Now $a(t)$ depends linearly upon past (upto t) values of $\nu(.)$ (or equivalently of $y(.)$). Clearly the best linear estimate of $a(t)$ given values of $y(.)$ upto t is $a(t)$ itself. However, the best linear estimate of $b(t)$ given values of $y(.)$ upto t is zero because the "future" values $\{\nu(\tau), t<\tau\leqslant t+\lambda\}$ are uncorrelated with $\{\nu(\tau), \tau \leqslant t\}$. Therefore,

$$\hat{y}(t+\lambda|t) = a(t) = \int_{-\infty}^{t} k(t+\lambda-\tau)\nu(\tau)\, d\tau \tag{17}$$

From this we can recognize the optimum filter $g(.)$ for calculating $\hat{y}(t+\lambda\,|\,t)$ from $\{\nu(\tau), \tau\leqslant t\}$

$$\hat{y}(t+\lambda|t) = \int_{0}^{t} g(t-\tau)\,\nu(\tau)\, d\tau$$

as

$$g(t) = \begin{cases} k(t+\lambda), & t\geqslant 0 \\ 0, & t<0. \end{cases} \tag{18}$$

In the frequency domain, this corresponds to

$$G(s) = \{\mathcal{L}[k(t)]\cdot e^{s\lambda}\}_+ = \{S_y^+(s)e^{s\lambda}\}_+ \tag{19}$$

The optimum filter for calculating $\hat{y}(t+\lambda|t)$ from $\{y(\tau), \tau\leqslant t\}$ is

$$H(s) = \frac{1}{S_y^+(s)}\cdot G(s) = \frac{1}{S_y^+(s)}\{S_y^+(s)e^{s\lambda}\}_+$$

which, of course, is the formula we found earlier in Sections 3.3. and 3.4.

The above discussion gives us a simple formula for the mean-square error:

$$E[y(t+\lambda)-\hat{y}(t+\lambda|t)]^2 = E[b^2(t)]$$

$$= E\int_{t}^{t+\lambda}\int_{t}^{t+\lambda} k(t+\lambda-\tau_1)k(t+\lambda-\tau_2)\,\nu(\tau_1)\nu(\tau_2)\, d\tau_1\, d\tau_2$$

$$= \int_{t}^{t+\lambda} k^2(t+\lambda-\tau_2)d\tau_2 = \int_{0}^{\lambda} k^2(t)dt \tag{20}$$

Exercise 3.4.1.

Calculate $\hat{y}(t+\lambda|t)$ given observations $\{x(\tau), -\infty<\tau<t\}$ where

$$E\, y(t)y(\tau) = R_y(t,\tau) = p(t)p(\tau)\exp -\alpha|t-\tau|$$

and the values of the function p(.) always lie between the levels 1 and 2.

Exercise 3.4.2. Scalar Wide-Sense Markov Processes

Let y(.) be a process with covariance function

$$R(t,s) = f(\min(t,s))g(\max(t,s))$$

where

$$g(t) \neq 0 \quad \text{for all} \quad t \geqslant 0.$$

a) Show that the linear least-squares predicted estimate $\hat{y}(t+\lambda|\, t)$, $\lambda > 0$, depends only on y(t). This is known as the wide-sense Markov (w.s.m.) property.

b) Show that a process $y(\cdot)$ is wide-sense Markov if and only if its covariance has the form given above with the additional requirement that

$$0 \leqslant h(t) \leqslant h(s) \text{ for } t < s.$$

is a nonnegative nondecreasing function, i.e., $h(t) \leqslant h(s)$ for $t > s$.

c) Show that

$$R(t,s) = e^{-\alpha|t-s|}, \quad \alpha > 0.$$

satisfies these conditions, and furthermore that this is the only stationary covariance with the wide-sense Markov property.

Exercise 3.4.3. a) A (Wiener) process with covariance min(t,s) is the simplest non-stationary w.s.m. process. Show that any w.s.m. process can be transformed into a generalized Wiener process by the transformation

$$w_h(t) = y(t)/g(t)$$

with

$$Ew_h(t)w_h(s) = \min(h(t),h(s)) = h(\min(t,s)).$$

b) Show that any generalized Wiener process y(.) is a Wiener process with respect to a new time scale τ defined by

$$\tau = h(t),$$

i.e., $z(\tau) \triangleq y(h^{-1}(\tau))$ is a standard Wiener process; h(.) can be thought of as the natural "clock" for the generalized Wiener process. This transformation is quite useful in simplifying the study of various problems for scalar wide-sense Markov processes (see, e.g., Exercise 4.7).

Exercise 3.4.4.

Show that the process

$$y(t) = \int_0^t h(t)k(s)u(s)ds \ , \quad u(\cdot) = \text{white noise}$$

is wide-sense Markov and compute its covariance function.

Exercise 3.4.5.

Let a process y(.) be the solution of the differential equation

$$\frac{dy(t)}{dt} + \alpha y(t) = u(t) \, , \quad t \geq 0$$

$$E\, u(t)u(s) = \delta(t-s), \quad E\, u(t)\, y(0) = 0$$

Show that the best prediction of $y(t_0 + \lambda)$ given $\{y(\tau) \, , \, \tau \leq t_0\}$ is the solution at time λ of the differential equation

$$\frac{dm(t)}{dt} + \alpha m(t) = 0, \quad t \geq 0, \quad m(0) = y(t_0)$$

Exercise 3.4.6.

If y(.) obeys

$$\ddot{y}(t) + \sqrt{2}\, \dot{y}(\cdot) + y(t) = y(t), \quad t \geq 0$$

$$E\, u(t)u(s) = \delta(t-s) \, , \quad E\, u(t)y(0) = 0 = E\, u(t)\dot{y}(0)$$

show that $y(t_0 + \lambda \mid \{ y(\tau) \, , \, \tau \leq t_0 \})$ is the solution at time λ of the differential equation

$$\frac{d^2 m(t)}{dt^2} + \sqrt{2}\, \frac{dm(t)}{dt} + m(t) = 0 \, ,$$

$$m(0) = y(t_0) \, , \quad \dot{m}(0) = \dot{y}(t_0) \, .$$

3.5. Some Examples; State-space Models

In this subsection we shall discuss some examples in a rather leisurely fashion. Our aim is to shed more light on the material of Section 3.4 and to prepare the way for the Kalman filters to be studied soon.

Example 3.5.1 :

We wish to calculate $\hat{y}(t+\lambda \mid t)$, $\lambda > 0$, for a process y(.) with power spectral density

$$S_y(\omega) = \frac{1}{1+\omega^4} .$$

Solution: We shall present several methods.

Method I: We let $s = i\omega$ and form

$$S_y(s) = \frac{1}{1+s^4} = \frac{1}{(s^2+\sqrt{2}\,s+1)(s^2-\sqrt{2}\,s+1)}$$

Therefore

$$S_y^+(s) = \frac{1}{s^2+\sqrt{2}\,s+1}$$

The optimum filter is

$$H(s) = \frac{1}{S_y^+(s)} \left\{ S_y^+(s)e^{s\lambda} \right\}_+$$

Now

$$\left\{ \frac{e^{s\lambda}}{s^2+\sqrt{2}\,s+1} \right\}_+ = \left\{ \frac{e^{s\lambda}/\sqrt{2}\,i}{s+\frac{1}{\sqrt{2}}-\frac{i}{\sqrt{2}}} \right\}_+ - \left\{ \frac{e^{s\lambda}/\sqrt{2}\,i}{s+\frac{1}{\sqrt{2}}+\frac{i}{\sqrt{2}}} \right\}_-$$

$$= e^{-\lambda/\sqrt{2}} \frac{e^{i\lambda/\sqrt{2}}}{\sqrt{2}\,i\left(s+\frac{1}{\sqrt{2}}-\frac{i}{\sqrt{2}}\right)} - e^{-\lambda/\sqrt{2}} \frac{e^{-i\lambda/\sqrt{2}}}{\sqrt{2}\,i\left(s+\frac{1}{\sqrt{2}}+\frac{i}{\sqrt{2}}\right)}$$

$$= e^{-\lambda/\sqrt{2}} \frac{\left\{ \sqrt{2}\left(s+\frac{1}{\sqrt{2}}\right)\sin\frac{\lambda}{\sqrt{2}} + \cos\frac{\lambda}{\sqrt{2}} \right\}}{s^2+\sqrt{2}\,s+1}$$

Therefore

$$H(s) = A + B\,s ,$$

$$h(t) = A\,\delta(t) + B\,\delta^{(1)}(t)$$

and with $*$ denoting convolution,

$$\hat{y}(t+\lambda|t) = h(t) * y(t)$$
$$= Ay(t) + By^{(1)}(t)$$

where

$$A = e^{-\lambda/\sqrt{2}}\left(\cos\frac{\lambda}{\sqrt{2}} + \sin\frac{\lambda}{\sqrt{2}}\right)$$

$$B = \sqrt{2}\, e^{-\lambda/\sqrt{2}} \sin\frac{\lambda}{\sqrt{2}} .$$

Method II: The above is such a simple answer that we might ask if a more direct solution is not possible. To do this and also with an eye on future extensions to the nonstationary case, let us see what happens with the alternative time-domain method of Section 3.4.

We first determine the innovations representation. This is easy and the solution is shown by the following block diagram.

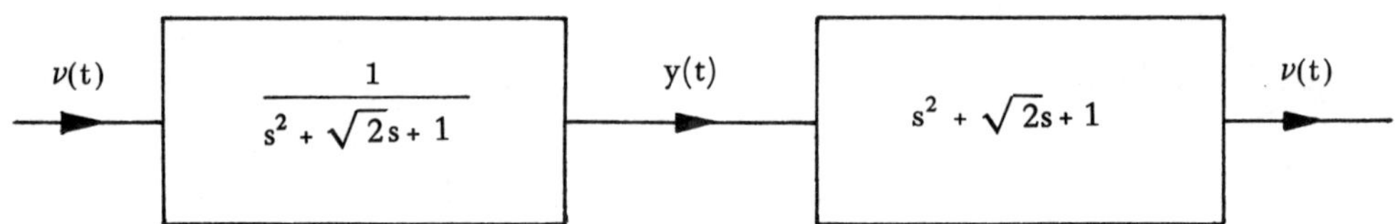

In the time domain

$$y(t) = \int_{-\infty}^{t} k(t-\tau)\, \nu(\tau)\, d\tau$$

where

$$k(t) = \sqrt{2}\, e^{-t/\sqrt{2}} \sin t/\sqrt{2}$$

and

$$\nu(t) = \ddot{y}(t) + \sqrt{2}\, \dot{y}(t) + y(t).$$

Now

$$y(t+\lambda) = \int_{-\infty}^{t} k(t+\lambda-\tau)\, \upsilon(\tau)\, d\tau + \int_{t}^{t+\lambda} k(t+\lambda-\tau)\upsilon(\tau)\, d\tau,$$

and since future $\upsilon(.)$ are uncorrelated with past $\upsilon(.)$, and hence past y(.), we have

$$\hat{y}(t+\lambda|t) \triangleq \hat{y}(t+\lambda) = \int_{-\infty}^{t} k(t+\lambda-\tau)\, \upsilon(\tau)\, d\tau$$

$$= 1(t)k(t+\lambda) * \upsilon(t)$$

$$= 1(t)k(t+\lambda) * (\ddot{y}(t) + \sqrt{2}\, \dot{y}(t) + y(t))$$

We can now use the relations,

$$\dot{y}(t) * 1(t)k(t+\lambda) = y(t) * \underbrace{\delta^{(1)}(t) * 1(t)k(t+\lambda)}$$

$$= y(t) * \delta(t)k(\lambda) + y(t) * 1(t)\dot{k}(t+\lambda)$$

$$\ddot{y}(t) * 1(t)k(t+\lambda) = y(t) * [\delta^{(1)}(t)k(\lambda) + \delta(t)k(\lambda) + 1(t)\ddot{k}(t+\lambda)] .$$

to obtain $\hat{y}(t+\lambda)$ in terms of $y(t)$ and $\dot{y}(t)$. The algebra is considerable however, and it is better to work in the frequency domain. Note first that

$$\hat{y}(t+\lambda) = \int_{-\infty}^{t} \sqrt{2}\, e^{-(t+\lambda-\tau)/\sqrt{2}} \sin\left(\frac{t+\lambda-\tau}{\sqrt{2}}\right) v(\tau)\, d\tau$$

$$= \sqrt{2}\, e^{-\lambda/\sqrt{2}} \cdot \int_{-\infty}^{t} e^{-(t-\tau)/\sqrt{2}} \left(\sin\frac{(t-\tau)}{\sqrt{2}} \cos\frac{\lambda}{\sqrt{2}} + \right.$$

$$\left. + \sin\frac{\lambda}{\sqrt{2}} \cos\frac{(t-\tau)}{\sqrt{2}}\right) v(\tau)\, d\tau$$

which shows that the transfer function from $v(.)$ to $\hat{y}(. + \lambda)$ is

$$\sqrt{2}\, e^{-\lambda/\sqrt{2}} \left[\cos\frac{\lambda}{\sqrt{2}} \cdot \frac{1/\sqrt{2}}{s^2 + \sqrt{2}\, s + 1} + \sin\frac{\lambda}{\sqrt{2}} \cdot \frac{s + 1/\sqrt{2}}{s^2 + \sqrt{2}\, s + 1}\right]$$

The transfer function from $y(.)$ to $v(.)$ is

$$s^2 + \sqrt{2}\, s + 1$$

so that the overall transfer function from $y(.)$ to $y(. + \lambda)$ is of the form $A + Bs$, where

$$A = e^{-\lambda/\sqrt{2}} \left(\cos\frac{\lambda}{\sqrt{2}} + \sin\frac{\lambda}{\sqrt{2}}\right)$$

$$B = \sqrt{2}\, e^{-\lambda/\sqrt{2}} \sin\frac{\lambda}{\sqrt{2}} ,$$

as previously obtained in Method I.

Method III: State Equations.

The time-domain method just described is conceptually simple, but involves a lot of algebra if we insist on working completely in the time-domain, as we should if we wish to obtain some ideas for the nonstationary problem. It turns out that the use of state equations enables a simple time-domain proof, the ideas of which can in fact be taken over to the nonstationary case.

For this we write a state-variable model for the innovations representation as

$$\begin{bmatrix} \dot{x}_1(t) \\ \dot{x}_2(t) \end{bmatrix} = \underbrace{\begin{bmatrix} -\sqrt{2} & -1 \\ 1 & 0 \end{bmatrix}}_{A} \begin{bmatrix} x_1(t) \\ x_2(t) \end{bmatrix} + \underbrace{\begin{bmatrix} 1 \\ 0 \end{bmatrix}}_{b} \upsilon(t)$$

$$y(t) = \underbrace{[0 \quad 1]}_{c} \begin{bmatrix} x_1(t) \\ x_2(t) \end{bmatrix} = x_2(t)$$

This is the so-called controller canonical form for the transfer function

$$S_y^+(s) = \frac{1}{s^2 + \sqrt{2}\, s + 1}$$

Note that

$$\dot{x}_2(t) = x_1(t)$$

so that

$$x_1(t) = \dot{y}(t)$$
$$x_2(t) = y(t)$$

Now in matrix notation we have

$$\dot{x}(t) = A\, x(t) + bu(t)$$
$$y(t) = cx(t)$$

and to find $\hat{y}(t + \lambda \mid t)$ we write

$$\hat{y}(t + \lambda \mid t) = c\hat{x}(t + \lambda \mid t)$$

But

$$x(t + \lambda) = e^{A(t + \lambda - t)} x(t) + \int_t^{t+\lambda} e^{A(t + \lambda - \tau)}\, b\upsilon(\tau)\, d\tau$$

Therefore

$$\hat{x}(t + \lambda \mid t) = e^{A\lambda} \hat{x}(t \mid t)$$

where

$$\hat{x}(t \mid t) = \text{the 1.1.s.e. of } x(t) \text{ given } \{y(\tau), \tau \leqslant t\}.$$

But

$$\dot{y}(\tau) = x_1(\tau), \quad y(\tau) = x_2(\tau) ,$$

so that given $\{y(\tau), \tau \leqslant t\}$ we know $x_1(t)$ and $x_2(t)$. Therefore

$$\hat{x}(t|t) = x(t)$$

and

$$\hat{y}(t+\lambda|t) = ce^{A\lambda}x(t)$$

But some simple algebra shows that

$$e^{A\lambda} = \left[\mathcal{L}^{-1}(sI-A)^{-1}\right]_{t=\lambda}$$

$$= e^{-\lambda/\sqrt{2}} \begin{bmatrix} \cos\frac{\lambda}{\sqrt{2}} - \sin\frac{\lambda}{\sqrt{2}} & -\sqrt{2}\sin\frac{\lambda}{\sqrt{2}} \\ \sqrt{2}\sin\frac{\lambda}{\sqrt{2}} & \cos\frac{\lambda}{\sqrt{2}} + \sin\frac{\lambda}{\sqrt{2}} \end{bmatrix},$$

Therefore

$$ce^{A\lambda}x(t) = \sqrt{2}e^{-\lambda/\sqrt{2}}\sin\frac{\lambda}{\sqrt{2}}\,\dot{y}(t) + e^{-\lambda/\sqrt{2}}\left(\cos\frac{\lambda}{\sqrt{2}} + \sin\frac{\lambda}{\sqrt{2}}\right)y(t).$$

The use of the state-variable description involves some lesser algebra, or to put it more correctly, it provides a more compact notation in which much of the algebra is hidden (especially in the computation of $e^{A\lambda}$). The state-variable method also gives the formula for $\hat{y}(t+\lambda|t)$ directly rather than just giving the impulse response of the optimum filter.

Example 3.5.2.

Now let

$$S_y(s) = \frac{1 - s^2}{1 + s^4}$$

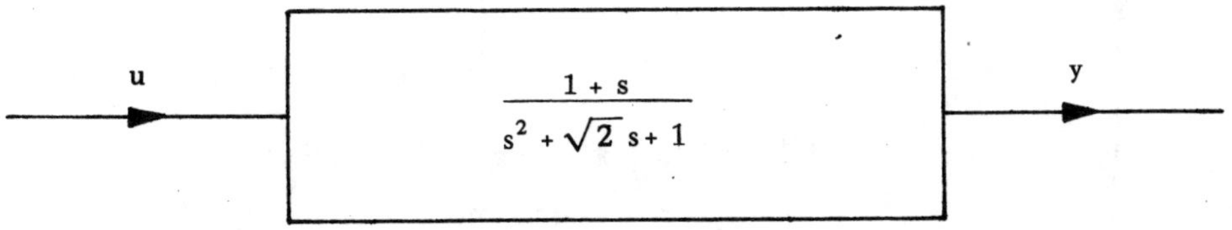

u → $\dfrac{1+s}{s^2+\sqrt{2}\,s+1}$ → y

$$\begin{bmatrix} \dot{x}_1 \\ \dot{x}_2 \end{bmatrix} = \begin{bmatrix} -\sqrt{2} & -1 \\ 1 & 0 \end{bmatrix} \begin{bmatrix} x_1 \\ x_2 \end{bmatrix} + \begin{bmatrix} 1 \\ 0 \end{bmatrix} u$$

$$y = [1 \quad 1]\,\mathbf{x} = x_1 + x_2 \,.$$

$$x(t+\lambda) = e^{A\lambda}x(t) + \int_t^{t+\lambda} e^{A(t-\tau)}bu(\tau)\, d\tau$$

$$\hat{x}(t+\lambda|t) = e^{A\lambda}\,\hat{x}(t|t)\,.$$

But now $x_1(t)$ and $x_2(t)$ are not determined directly from $y(t)$; only their sum is. Therefore we have to calculate $\hat{x}(t|t)$. However note that

$$\dot{x}_2(t) = x_1(t)$$

so that

$$y(t) = \dot{x}_2(t) + x_2(t)$$

or

$$x_2(t) = \int_{-\infty}^{t} e^{-(t-\tau)}y(\tau)\, d\tau = \hat{x}_2(t)$$

Therefore

$$x_1(t) = y(t) - \int_{-\infty}^{t} e^{-(t-\tau)}y(\tau)\, d\tau = \hat{x}_1(t)\,.$$

$$\hat{y}(t+\lambda) = ce^{A\lambda} \begin{bmatrix} y(t) - e^{-t} * y(t) \\ e^{-t} * y(t) \end{bmatrix}$$

$$= \left(e^{-\lambda/\sqrt{2}} \cos\frac{\lambda}{\sqrt{2}} + (\sqrt{2}-1)\, e^{-\lambda/\sqrt{2}} \sin\frac{\lambda}{\sqrt{2}} \right) y(t)$$

$$-\, 2(\sqrt{2}-1)\, e^{-\lambda/\sqrt{2}} \sin\frac{\lambda}{\sqrt{2}} \int_{-\infty}^{t} e^{-(t-\tau)}y(\tau)\, d\tau\,.$$

An Alternative Method

Things will be much easier if we use the observer canonical realization :

$$\begin{bmatrix} \dot{x}_1 \\ \dot{x}_2 \end{bmatrix} = \begin{bmatrix} -\sqrt{2} & 1 \\ -1 & 0 \end{bmatrix} \begin{bmatrix} x_1 \\ x_2 \end{bmatrix} + \begin{bmatrix} 1 \\ 1 \end{bmatrix} u$$

$$y = [1 \quad 0]\,\mathbf{x} = x_1$$

Then

$$x_1 = y$$

and

$$\dot{x}_2 = -x_1 + u .$$

$$\dot{x}_1 = -\sqrt{2}\, x_1 + x_2 + u ,$$

so that

$$u = \dot{y} + \sqrt{2}\, y - x_2 .$$

Therefore

$$\dot{x}_2 = -y + \sqrt{2}\, y + \dot{y} - x_2$$

or

$$\dot{x}_2 + x_2 = \dot{y} + (\sqrt{2} - 1)\, y ,$$

and

$$x_2(t) = e^{-t} * \dot{y} + (\sqrt{2} - 1)\, e^{-t} * y$$
$$= (\delta - e^{-t}) * y + (\sqrt{2} - 1)\, e^{-t} * y .$$

The point is that there are many state-space models for a given transfer function. [And there can be many transfer functions that will yield the "same" output process (i.e., a process with the same mean and covariance functions) when driven by a (unit-intensity) white noise input process.]

All these possible different models will have different mean-square-errors for the different state-estimates, but the mean-square-error in $y(t)$ ($E\,[y(t) - \hat{y}(t)]^2$, $\hat{y}(t) = C_\alpha \hat{x}_\alpha(t)$, $\alpha = 1,2,\ldots$ corresponding to the different state models) will of course be the same.

Example 3.5.3. Time-Variant State-Space Models

An important advantage of the state-space models is that conceptually and numerically (if computers are used to solve the equations), there is not much difference between time-invariant and time-variant models.

Thus, if we had a constant parameter equation

$$\dot{x}(t) = A\,x(t) + bu(t)$$

a way (a naive one, but sufficient to make the point) of solving it on a computer is to approximate it by the difference equations

$$x(t+\Delta) = (I + A\Delta)x(t) + \Delta\, bu(t)\,, \quad t = 0, \Delta, 2\Delta, \ldots$$

Starting with $x(0) = x_0$, this difference equation gives us successively $x(\Delta), x(2\Delta), \ldots$; in this calculation it makes little difference whether $\{A, b\}$ are constant or change with time in some known way. In other words, we might as easily consider state-space equations with time-variant coefficients,

$$\dot{x}(t) = A(t)x(t) + b(t)u(t), \quad t \geqslant t_0$$
$$z(t) = c(t)x(t),$$

where $u(.)$ is a white-noise process, uncorrelated with the (random) initial condition $x(t_0)$. Then

$$z(t+\lambda) = c(t+\lambda)\Phi(t+\lambda,t)x(t) + \int_t^{t+\lambda} c(t+\lambda)\Phi(t+\lambda,\tau)b(\tau)u(\tau)\, d\tau$$

where $\Phi(.\,,.)$ is the state-transition matrix of $A(.)$, i.e., it is the unique solution of the linear matrix equation

$$\frac{d\Phi(t,\tau)}{dt} = A(t)\Phi(t,\tau), \quad \Phi(\tau,\tau) = I\,.$$

For constant $A(.)$ matrices

$$\Phi(t,\tau) = \exp A(t-\tau)\,.$$

In the time variant case, $\Phi(.,.)$ is hard to find explicitly, but it serves as a convenient notational device (see [7, Ch. 9]).

We see that

$$\hat{z}(t+\lambda|t) = c(t+\lambda)\Phi(t+\lambda,t)\hat{x}(t|t)$$

where

$$\hat{x}(t|t) = \text{the l.l.s.e. of } x(t) \text{ given past data, } \{z(\tau),\ \tau \leq t\}$$

Now if the state-space model is causally invertible, we can reconstruct $\{u(\tau), \tau \leq t, x(t_0)\}$ from $\{z(\tau), \tau \leq t\}$ and thereby also reconstruct $\{x(\tau), \tau \leq t\}$ by solving the state-equation forwards.

If the model is not causally invertible, then all we can say is that the predicted estimates $\hat{x}(t + \lambda | t)$ and $\hat{z}(t + \lambda | t)$ are determined by the filtered estimate $\hat{x}(t|t)$. Since the notion of power-spectrum is no longer available (because of the time-variant coefficients and/or because $t_0 \neq -\infty$), we shall have to look for other means of determining $\hat{x}(t|t)$. This question will be discussed in Sections 4-6.

Example 3.5.4. Vector Processes

Another feature of the use of state-space models is that there is no particular conceptual difference in handling multi-input multi-output models

$$\dot{x}(t) = A(t)x(t) + B(t)u(t), \quad t \geq t_0$$
$$z(t) = C(t)x(t),$$

where C(.) is say a p x n matrix, A(.) is n x n and B(.) is n x m. When z(.) is a stationary process observed from $t_0 = -\infty$, the Wiener-Hopf method will require the factorization of a spectral-density **matrix**. This is more difficult, though various methods are available. A set of references can be found in [8, pp. 167-168]. While this factorization problem does not explicitly arise when state-models are used (as we shall see in Section 6), clearly one must be doing something closely related, a question we shall discuss in Sections 7 and 8.

At this point, then, it seems appropriate to turn in the next section to the early efforts that were made to generalize the Wiener theory.

REFERENCES

[1] N. Wiener, **Extrapolation, Interpolation and Smoothing of Stationary Time Series, with Engineering Applications.** New York: Technology Press and Wiley, 1949, (Originally issued in February 1942, as a classified Nat. Defense Res. Council Rep).

[2] M.C. Yovits and J.L. Jackson, "Linear filter optimization with game theory considerations", in **IRE Nat. Conv. Rec.**, pt. 4, pp. 193-199, 1955.

[3] N. Levinson, "A heuristic exposition of Wiener's mathematical theory of prediction and filtering", **J. Math. Phys.**, vol. 15, pp. 110-119, July 1947; reprinted as an Appendix in [1].

[4] M.G. Krein, "Integral equations on a half -axis with kernel depending on the difference of the arguments", **Usp. Math. Nauk.**, vol. 13, pp. 3-120, 1958, Amer. Math. Socy. Transl.

[5] H.W. Bode and C.E. Shannon, "A simplified derivation of linear least squares smoothing and prediction theory", **Proc. IRE**, vol. 38, pp. 417-425, April 1950.

[6] L.A. Zadeh and J.R. Ragazzini, "An extension of Wiener's theory of prediction", J. Appl. Phys., vol. 21, pp. 645-655, July 1950.

[7] T. Kailath, *Linear Systems,* New Jersey: Prentice-Hall, 1980.

[8] T. Kailath, "A view of three decades of linear filtering theory", **IEEE Trans. on Information Theory**, vol. IT-20, No. 2, pp. 145-181, March 1974.

4. GENERALIZATIONS OF WIENER FILTERING †

Soon after Wiener's work was understood, attempts were made to remove the restrictions of

i) semi-infinite observation intervals $(-\infty, t)$,

ii) stationary and jointly stationary signal and observation processes,

iii) scalar signal and observation processes. There is no difficulty in obtaining the appropriate integral equation to calculate

$$\hat{x}(t|T) \triangleq \text{the l.l.s.e. of } x(t) \text{ given } \{y(\tau), 0 \leq \tau \leq T\},$$

For even if none of the above restrictions is imposed, we can write

$$\hat{x}(t|T) = \int_0^T h(t, s)y(s)\, ds.$$

Then the projection theorem

$$x(t) - \hat{x}(t|T) \perp \{y(\tau), 0 \leq \tau \leq T\}$$

yields

$$R_{xy}(t, \tau) = \int_0^T h(t, s)R_{yy}(s, \tau)\, ds, \quad 0 \leq \tau \leq T. \tag{1}$$

If $T = t$, $\hat{x}(t|t) = \hat{x}(t)$ is called a **causal** or **filtered** estimate.
If $T > t$, $\hat{x}(t|T)$ is a **noncausal** or **smoothed** estimate.
If $t > T$, $\hat{x}(t|T)$ is a **predicted** estimate.

The integral equation (1) is known as a **Fredholm equation of the first kind** (Smithies [1]), (Cochran [2]) and is usually quite difficult to solve, either analytically or numerically. When y(.) is a scalar stationary process with a rational spectral density, i.e.,

$$S_y(\omega) = \frac{\omega^{2m} + b_{2m-2}\omega^{2m-2} + \ldots + b_0}{\omega^{2n} + a_{2n-2}\omega^{2n-2} + \ldots + a_0} \tag{2}$$

or

$$R_y(t) = \sum_{i=1}^{n} c_i e^{-d_i|t|} \tag{3}^*$$

† See also Appendix I.

(*) $\{a_i\}$, $\{b_i\}$ and $\{c_i\}$ must be real. For simplicity, we have also assumed that the roots $\{\pm d_i\}$ of the denominator in (2) are distinct; otherwise we shall have to include terms of the form $t^k e^{-d_i t}$ in (3).

then some explicit methods of solution are known – see, e.g., the text-books of Laning and Battin [3], Van Trees [4] and Whittle [5] .

For nonstationary "lumped" processes, i.e., processes generated by passing white-noise through lumped linear time-variant networks, the covariance has the form

$$R_y(t, s) = \begin{cases} \sum_1^n a_i(t)b_i(s) & t \geqslant s \\ \sum_1^n a_i(s)b_i(t), & t \leqslant s \end{cases} \tag{4}$$

or more compactly

$$R_y(t, s) = \begin{cases} A(t)B(s) & , \ t \geqslant s \\ B'(t)A'(s) & , \ t \leqslant s \end{cases} \tag{5}$$

where A(.) and B(.) are matrices of appropriate dimension. Perhaps the most general results were given by Shinbrot [6] for scalar processes – see also the books of J. Bendat [7] and of E.L. Peterson [8].

In many problems, the observation process contains some white noise, so that $R_y(t,s)$ has the form

$$R_y(t, s) = R(t)\delta(t-s) + K_0(t, s), \qquad R(t) > 0 \tag{6}$$

Since $R(t) > 0$, we can "normalize" the equation so that we have the form

$$R_{yy}(t, s) = \mathbf{I}\delta(t-s) + K(t, s) \tag{7}$$

Now Eq. (1) becomes

$$h(t, \tau) + \int_0^T h(t, s)K(s, \tau)\, ds = R_{xy}(t, \tau), \quad 0 \leqslant \tau \leqslant T \tag{8}$$

which is a **Fredholm equation of the second kind.** Such equations are much easier to solve, at least numerically, than are equations of the first kind, but analytical solutions are still essentially known only for K(t,s) of the form (5) [which, of course, includes (3)]. [The point is that for equations as in (8), the solution h(. , .) has the same smoothness properties as the given functions K(. , .) and $R_{xy}(\cdot,\cdot)$, while for equations of the first kind the solution h(.,.) can be much rougher – cf. Example 2.2.1].

Even when analytical solutions are known for (1) and (8), they are quite complicated, especially for vector processes. They also have the undesirable feature that the whole solution has to be modified if the observation interval is increased, as happens often in trajectory estimation problems. Thus in the late 50's, pressure developed for more effective "computational algorithms" as opposed to explicit "closed-form" solutions.

In discrete-time problems, such algorithms had already been developed by Gauss in his calculations of planetary orbits. The modern period was launched by the papers of Swerling [9] and especially Kalman, Bucy, and Stratonovich [10]-[12].

The latter work was characterized by a new assumption, viz., knowledge of the covariance function was replaced by knowledge of a "lumped" finite-dimensional model generating the signal process. That is, instead of the specification

$$y = z + v, \quad R_y = \mathbf{I}\delta(t-s) + A(t)B(s)1(t-s) + B'(t)A'(s)1(s-t) \tag{9}$$

we have a state-space model for the process (cf. Sec. 3.5)

$$\dot{\mathbf{x}}(t) = \mathbf{F}(t)\mathbf{x}(t) + \mathbf{G}(t)\mathbf{u}(t), \quad x(t_0) = x_0 \tag{10a}$$

$$\mathbf{z}(t) = \mathbf{H}(t)\mathbf{x}(t) \tag{10b}$$

$$\mathbf{y}(t) = \mathbf{H}(t)\mathbf{x}(t) + \mathbf{v}(t) \tag{10c}$$

$$E\, u_k = E\, v_k = E\, \mathbf{x}_0 = 0, \quad E\, \mathbf{x}_0\mathbf{x}_0' = \mathbf{\Pi}_0, \quad E\, \mathbf{u}(t) = E\, \mathbf{v}(t) = E\, \mathbf{x}_0 = 0, \tag{10d}$$

$$E\, \mathbf{u}(t)\mathbf{u}'(s) = \mathbf{Q}(t)\delta(t-s), \quad E\, \mathbf{v}(t)\mathbf{v}'(s) = \mathbf{I}\,\delta(t-s) \tag{10e}$$

$$E\, \mathbf{u}(t)\mathbf{v}'(s) = \mathbf{C}(t)\delta(t-s)\,. \tag{10f}$$

We assume that

$$\mathbf{u}(\cdot) \text{ is } m \times 1, \quad \mathbf{x}(\cdot) \text{ is } n \times 1, \quad \mathbf{z}(\cdot) \text{ is } p \times 1 \tag{10g}$$

and that matrices $\mathbf{F}(.)$, $\mathbf{G}(.)$, $\mathbf{H}(.)$, $\Pi(.)$, $\mathbf{Q}(.)$, $\mathbf{C}(.)$ are assumed to be known a priori along with their dimensions. Henceforth for convenience we shall not use any explicit (underscoring) notation for vectors or matrices. Clearly knowledge of the model (10) completely determines the covariance function, though of course there are many models that can give rise to the same covariance function.

Covariance Functions of Lumped Models

Some fairly straightforward calculations show that for $y(.)$ as in (10) the covariance function is

(11a) $$R_y(t, s) = I\delta(t - s) + K(t, s)$$

where if $z(t) = H(t)x(t)$, then

(11b) $$K(t, s) = E\ z(t)z'(s) + E\ v(t)z'(s) + E\ z(t)v'(s)$$

(11c) $$= M(t)\Phi(t, s)N(s)1(t-s) + N'(t)\Phi'(s, t)M'(s)1(s-t)$$

and $\Phi(t,s)$ is the state-transition matrix defined by the equation

(11d) $$\frac{d\Phi(t, s)}{dt} = F(t)\Phi(t, s), \qquad \Phi(t_0, t_0) = I ,$$

and

(11e) $$M(t) = H(t) , \quad N(t) = \Pi(t)H'(t) + G(t)C(t) .$$

The matrix $\Pi(.)$ is the **variance matrix** of the state vector $x(.)$,

(12a) $$\Pi(t) = E\ x(t)x'(t)$$

It is important to note that when $x(.)$ obeys a vector linear equation of the form (10) then $\Pi(.)$ obeys the equation

(12b) $$\dot{\Pi}(t) = F(t)\Pi(t) + \Pi(t)F'(t) + G(t)Q(t)G'(t), \quad \Pi(0) = \Pi_0$$

The simple derivation of this very useful formula is left to the reader.

Comparison of (9) and (11)-(12) shows that we can make the (nonunique) identifications

(13a) $$A(t) = M(t)\Phi(t, t_0) ,$$

(13b) $$B(t) = \Phi(t_0, t)(\Pi(t)M'(t) + G(t)C(t)) .$$

Clearly there can be many sets of parameters $\{F(.), G(.), H(.), Q(.), C(.), \Pi_0\}$ yielding a given covariance function, i.e., essentially a given $A(.)$ and $B(.)$. However in many problems, e.g., in trajectory estimation, it is a model that is directly given or that is easier to calculate from the given data. We shall study such problems in Secs. 5, 6, and 7. In Section 8 we shall return to problems with covariance information.

Discrete-Time Lumped Models

Actually, discrete-time problems were the first to be completely studied from the viewpoint of recursive computation, beginning in fact with Gauss. A detailed historical account is given in [13]. We shall be primarily concerned with a state-variable model introduced by Kalman (1960), though we hasten to add that other models, especially the so-called mixed autoregressive-moving average models of time-series analysis, are also important and can be studied [13].

The model studied by Kalman was of the general form:

$$x_{k+1} = \Phi_{k+1,k} x_k + \Gamma_{k+1} u_k , \quad x_{k_0} = x_0 \tag{14a}$$

$$z_k = H_k x_k , \quad y_k = z_k + v_k \tag{14b}$$

$$E\, x_0 = 0, \quad E\, x_0 x_0' = \Pi_0, \quad E\, u_k x_0' \equiv 0 \tag{14c}$$

$$E\, u_k u_\ell' = Q_k \delta_{k\ell} , \quad E\, u_k v_\ell' = C_k \delta_{k\ell} \tag{14d}$$

$$E\, v_k v_\ell' = R_k \delta_{k\ell} . \tag{14e}$$

The matrices $\{\Phi, \Gamma, H, \Pi_0, Q, C, R\}$ are assumed to be known and it is assumed that Q_k is nonnegative definite while R_k is positive definite,

$$Q_k \geqslant 0, \qquad R_k > 0 .$$

We assume also that we have

$$m \text{ inputs } u, \; n \text{ states } x, \text{ and } p \text{ outputs } y . \tag{14f}$$

We can calculate the covariance function of a process of the above form as

$$R_y(k,\ell) = E\, y_k y_\ell' = \mathcal{R}_k \delta_{k\ell} + M_k \Phi_{k,\ell+1} N_\ell 1(k-\ell) + N_k' \Phi_{\ell,k+1}' M_\ell' 1(\ell - k) \tag{15a}$$

where

$$1(k-\ell) = \begin{cases} 1, & k > \ell \\ 0, & k \leqslant \ell \end{cases}$$

$$\mathcal{R}_k = R_k + H_k \Pi_k H_k'$$

$$M_k = H_k, \quad N_k = \Phi_{k+1,k} \Pi_k H_k' + \Gamma_{k+1} C_k \tag{15b}$$

and Π_k is the variance matrix of the state-vector x_k,

$$\Pi_k = E\, x_k x_k' . \tag{16a}$$

It is important to note that Π_k can be calculated via the difference equation

(16b) $$\Pi_{k+1} = \Phi_{k+1,k}\Pi_k\Phi'_{k+1,k} + \Gamma_{k+1}Q_k\Gamma'_{k+1}, \quad \Pi_{k_0} = \Pi_0$$

We observe again that there can be several different models corresponding to a given covariance function; however, in many problems a model may be directly specified. We shall study the filtering problem for such models in Sec. 5.

An Alternative Discrete-Time Model

Sometimes it is more natural to work with the model

(17a) $$x_{k+1} = \Phi_{k+1,k}x_k + \Gamma_{k+1}u_{k+1},$$

(17b) $$z_k = H_k x_k, \quad y_k = z_k + v_k$$

(17c) $$E\, x_0 = 0, \quad E\, x_0 x_0' = \Pi_0, \quad E\, u_k x_0' \equiv 0 \equiv E\, v_k x_0'$$

(17d) $$E\, u_k u_\ell' = Q_k \delta_{k\ell}, \quad E\, u_k v_\ell' \equiv C_k \delta_{k\ell}, \quad E\, v_k v_\ell' = R_k \delta_{k\ell}.$$

In this case, we will have

(18) $$R_y(k,\ell) = \mathcal{R}_k \delta_{k\ell} + M_k \Phi_{k,\ell} N_\ell 1(k-\ell) + N_k' \Phi_{\ell,k}' M_\ell' 1(\ell - k)$$

where

(19a) $$\mathcal{R}_k = R_k + H_k \Pi_k H_k' + H_k \Gamma_k C_k + C_k' \Gamma_k' H_k'$$

(19b) $$M_k = H_k$$

(19c) $$N_k = \Pi_k H_k' + \Gamma_k C_k.$$

Applications of this model in stochastic realization and spectral factorization problems can be found in [14]-[15].

As noted earlier in the next three sections we shall present a number of results for state-space models. Then using the hindsight gained from these studies we shall show how convenient recursive solutions can also be obtained for the generalized Wiener-Hopf-type equations studied by Shinbrot and others.

Exercise 4.1. Degenerate Kernels

Suppose K(t,s) is a degenerate kernel of the form

$$K(t, s) = \sum_1^n \lambda_i \phi_i(t)\phi_i(s), \qquad n < \infty .$$

where the $\phi_i(.)$ are orthonormal over (0,T). Show that we can write

$$h(t, \tau) = \sum_1^n \frac{\lambda_i}{1 + \lambda_i} \phi_i(t)\phi_i(\tau) .$$

The results of this exercise suggest a widely-used approximation technique for Fredholm equations of the second kind.

Exercise 4.2.

From equation (10a), we can write

$$x(t) = \Phi(t, s)x(s) + \int_s^t \Phi(t, \tau)G(\tau)u(\tau)\, d\tau .$$

Use this result to show that

$$E\ x(t)x'(s) = \Phi(t, s)\,\Pi(s), \quad \text{for } t \geqslant s$$

where $\Pi(s)$ is given by (12a)-(12b).

Exercise 4.3.

Try to establish (12) by using the formula

$$\frac{d}{dt} E\, x(t)x'(t) = E\, \dot{x}(t)x'(t) + E\, x(t)\dot{x}'(t).$$

Remark: This exercise shows that some care (or convention) must be used in manipulating nonlinear functionals of white noise. The exercise provides a nice motivating example for the Itô stochastic calculus (see [16]).

Exercise 4.4.

Consider a Wiener process and assume that the values $X(s) = x$, $X(t) = y$, $t > s$, are known. Show that the conditioned process is w.s.m. with mean

$$m(\tau) = x + \frac{\tau - s}{t - s}(y - x)$$

$$\text{Cov}\ x(\tau_1)x(\tau_2) = \frac{(\tau_1 - s)(t - \tau_2)}{t - s}, \quad \tau_1 < \tau_2 .$$

Exercise 4.5. Sliding Memory Filters

We wish to find the 1.1.s.e. of $x(t + \lambda)$ from observations of a random process $\{ y(\tau), t - T \leqslant \tau \leqslant t \}$. Find the integral equation that the optimum filter $h(t, \tau)$,

$$\hat{x}(t + \lambda) = \int_{t-T}^{t} h(t, \tau) y(\tau)\, d\tau$$

must satisfy. Show that if $x(.)$ and $y(.)$ are jointly stationary processes, then we can assume that $h(t, \tau)$ depends only on $(t - \tau)$.

Exercise 4.6. A Simple Integral Equation

i) Show that when

$$R(t,s) = \min(h(t),h(s)),$$

the integral equation

$$\int_0^T R(t, s) a(s)\, ds = m(t), \quad 0 \leqslant t \leqslant T$$

is satisfied by

$$a(t) = -\frac{d}{dt}\left[\frac{\dot{m}(t)}{\dot{h}(t)}\right] + \left[\frac{m(0)}{h(0)} - \frac{\dot{m}(0)}{\dot{h}(0)}\right] \delta(t) + \frac{\dot{m}(T)}{\dot{h}(T)} \delta(t - T).$$

ii) Use the transformation to a generalized Wiener process to solve the integral equation when

$$R(t,s) = \exp - \alpha|t - s|.$$

REFERENCES

[1] F. Smithies, **Integral Equations,** New York: Cambridge University Press, 1962.

[2] J.A. Cochran, **Analysis of Linear Integral Equations,** New York: McGraw-Hill, 1972.

[3] H. Laning and R. Battin, **Random Processes in Automatic Control,** New York; McGraw-Hill, 1958.

[4] H.L. Van Trees, **Detection, Estimation, and Modulation Theory, Part I,** New York: Wiley, 1968.

[5] P. Whittle, **Prediction and Regulation,** New York: Van Nostrand Reinhold, 1963.

[6] M. Shinbrot, "A generalization of a method for the solution of the integral equation arising in optimization of time-varying linear systems with nonstationary inputs", **IRE Trans. Inform. Theory**, vol. IT-3, pp.220-225, Dec. 1957.

[7] J. Bendat, **Principles and Applications of Random Noise Theory,** New York: Wiley, 1958.

[8] E.L. Peterson, **Statistical Analysis and Optimization of Systems,** New York: Wiley, 1961.

[9] P. Swerling, "First-order error propagation in a stagewise smoothing procedure for satellite observations", **J. Astronaut. Sci.**, vol. 6, pp. 46-52, Autumn 1959; see also "A proposed stagewise differential correction procedure for satellite tracking and prediction", RAND Corp. Rep. p.1292, Jan. 1958.

[10] R.E. Kalman, "A new approach to linear filtering and prediction problems", **J. Basic Eng.**, vol. 82, pp. 34-45, March 1960.

[11] R.E. Kalman and R.S. Bucy, "New results in linear filtering and prediction theory", **Trans. ASME, Ser. D, J. Basic Eng.**, vol. 83, pp. 95-107, Dec. 1961.

[12] R.L. Stratonovich, "Application of the theory of Markoff processes in optimal signal detection", **Radio Eng. Electron. Phys.**, vol. 1, pp. 1-19, 1960 (Translation from Russian).

[13] T. Kailath, "A View of Three Decades of Linear Filtering Theory," **IEEE Trans. Inform. Theory**, vol. IT-20, pp. 146-181, 1974.

[14] J. Rissanen and T. Kailath, "Partial realization of stochastic processes", **Automatica,** vol. 8, pp. 380-386, July 1972.

[15] M. Gevers and T. Kailath, "An innovations approach to least-squares estimation – Part VI: Discrete-time innovations representations and recursive estimation", **IEEE Trans. Automat. Contr.**, vol. AC-18, pp. 588-600, Dec. 1973.

[16] T. Kailath and P. Frost, "Mathematical modeling for stochastic processes" in **Stochastic Problems in Control", (Proc. Symp. AACC),** Amer. Soc. Mech. Engrs., 1968.

5. DISCRETE–TIME RECURSIVE ESTIMATION AND THE KALMAN FILTER

We shall approach the concept of recursive least-squares estimates by considering the following problem. Suppose we have found

$$\hat{x}|_{k-1} = \text{the llse of } x \text{ given } \{y_0, \ldots, y_{k-1}\}$$

and that now we have an additional observed random variable y_k. What is a convenient way of exploiting our knowledge of $\hat{x}|_{k-1}$ to find $\hat{x}|_k$, without just redoing the whole problem with the data $\{y_0, \ldots, y_{k-1}, y_k\}$? An answer to this question can be given in several ways, going back at least to Gauss in the early nineteenth century. One approach is based on the notion of **innovations** (or new information), which was perhaps first used in this problem by Wold (1938) and Kolmogorov (1939).

Let

$$\epsilon_k = y_k - \hat{y}_{k\,|\,k-1}$$

Then this quantity is clearly uncorrelated with all past random variables $\{y_0, \ldots, y_{k-1}\}$ and may be regarded as a measure of the new information or the **innovation** in the random variable y_k. The point is that the new observation y_k does not itself bring us completely new information, since a part of it (namely $\hat{y}_{k\,|\,k-1}$) is already completely determined by the previous observations $\{y_0, \ldots, y_{k-1}\}$; the error $\epsilon_k = \tilde{y}_{k|k-1} = y_k - \hat{y}_{k\,|\,k-1}$ is really the part of y_k that is "new".

This interpretation immediately suggests that

$$\begin{aligned} \hat{x}|_k &= \hat{x}|_{k-1} + (\text{llse of } x \text{ given } \epsilon_k) \\ &= \hat{x}|_{k-1} + (E\, x\epsilon_k')(E\, \epsilon_k\epsilon_k')^{-1}\epsilon_k \end{aligned} \tag{1}$$

This is easily verified by first noting that

$$\tilde{x}|_k = x - \hat{x}|_k = \tilde{x}|_{k-1} - (E\, x\epsilon_k')(E\, \epsilon_k\epsilon_k')^{-1}\epsilon_k$$

is obviously orthogonal to $\{y_0, \ldots, y_{k-1}\}$. Next observe that

$$E\,\tilde{x}|_{k-1}y_k' = E\,\tilde{x}|_{k-1}(\epsilon_k' + \hat{y}_{k\,|\,k-1}')$$

$$= E\,\tilde{x}_{|k-1}\epsilon_k' = E\,x\epsilon_k' - E\,\hat{x}_{|k-1}\epsilon_k'$$
$$= E\,x\epsilon_k'.$$

Also

$$E\ \epsilon_k y_k' = E\ \epsilon_k\epsilon_k' + E\ \epsilon_k\hat{y}_{k|k-1}$$
$$= E\ \epsilon_k\epsilon_k'\ .$$

Therefore

$$E\,\tilde{x}_{|k}y_k' = (E\,\tilde{x}_{|k-1}y_k') - (E\,x\epsilon_k')\,(E\,\epsilon_k\epsilon_k')^{-1}\ E\,\epsilon_k y_k'$$
$$= (E\,x\epsilon_k') - (E\,x\epsilon_k') = 0\,.$$

In other words, $\tilde{x}_{|k}$ is orthogonal to $\{y_0, \ldots, y_k\}$ and therefore by the projection theorem $\tilde{x}_{|k}$ as given by (1) must be the linear least-squares estimate of x given $\{y_0, \ldots, y_k\}$.

Another derivation of this result will be instructive. Let

$$\epsilon_j = y_j - \hat{y}_{j|j-1}\ ,\ j = 0, 1, \ldots$$

with $\hat{y}_{0|-1} = 0$. Then direct calculation, or analogy with the Gram-Schmidt procedure, will show that the $\{\epsilon_j\}$ are uncorrelated with each other and therefore form a white-noise innovations process. It is also easy to see that (provided there are no linear dependencies between the variables $\{y_j\}$) the innovations $\{\epsilon_j\ ,\ j \leqslant i\ ,\ i = 1, 2,\}$ and the observations $\{y_j, j \leqslant i\}$ can each be obtained from the other by causal and causally invertible linear transformations. Therefore one set can be replaced by the other without loss of information, and we can write

$$\hat{x}_{|k} = \text{the llse of } x \text{ given } \{y_0, \ldots, y_k\}$$

$$= \text{the llse of } x \text{ given } \{\epsilon_0, \ldots, \epsilon_k\}$$

But since the $\{\epsilon_i\}$ are uncorrelated with each other, it follows easily that

$$\hat{x}_{|k} = \sum_{i=0}^{k} (E\ x\epsilon_i')(E\ \epsilon_i\epsilon_i')^{-1}\epsilon_i \qquad (2)$$

$$= \hat{x}_{|k-1} + (E\, x\epsilon'_k)(E\, \epsilon_k \epsilon'_k)^{-1} \epsilon_k ,$$

the same result (1) as obtained before.

The simple formulas (1) and (2) are the basis for all recursive estimation formulas.

Estimation of a Random Process $\{x_k\}$

Thus suppose that, instead of estimating just a single random variable, we wish to estimate the values of a stochastic process $\{x_\ell\}$ given $\{y_k\}$. Then we have to ask for

$$\hat{x}_{\ell|k} = \text{the 11se of } x_\ell \text{ given } \{y_0, \ldots, y_k\}$$

Now as k increases, ℓ can be specified in different ways :

$\ell = k + 1$: **one-step predicted** estimate
$\ell = k$: **filtered** estimate
$\ell = k - \Delta$, Δ fixed : **fixed-lag smoothed** estimate (with lag Δ)
ℓ fixed : **fixed point smoothed** estimate

If k is fixed and ℓ increases from 0 to k, we say that we are finding **fixed-interval** smoothed estimates.

These different estimates all can be obtained by setting $x = x_\ell$, in (1), with ℓ appropriately specified for each kind of estimate ; $E\, x_\ell \epsilon'_k$ will have to be determined from the given statistical information about the $\{x_\ell\}$ and $\{y_k\}$ processes. Now, depending upon the form of this statistical information, i.e., upon any special features it may have, we may be able to elaborate upon the basic formula (1) to some degree.

Data Linearly Related to the Unknown

For example, we may be told that

$$\begin{aligned} R_{yy}(\ell, k) &\triangleq E\, y(\ell) y'(k) \\ &= H_\ell R_{xy}(\ell,k) + R_k \delta_{k\ell} \end{aligned} \tag{3}$$

where $\{H_k\}$ is a known sequence of matrices.

A closely related specification is that

(4) $$y_k = H_k x_k + v_k \,, \quad E\, v_k v_\ell' = R_k \delta_{k\ell} \quad | \quad E\, x_k v_\ell' \equiv 0.$$

This specification (4) gives the relations

$$\epsilon_k = y_k - \hat{y}_{k|k-1}$$

(5) $$= y_k - H_k \hat{x}_{k|k-1} = H_k \tilde{x}_{k|k-1} + v_k \,,$$

and

(6) $$E\, \epsilon_k \epsilon_k' = R_k^{\epsilon} = H_k P_{k|k-1} H_k' + R_k$$

where

$$P_{k|k-1} \triangleq E\, \tilde{x}_{k|k-1} \tilde{x}_{k|k-1}' \,, \quad \tilde{x}_{k|k-1} = x_k - \hat{x}_{k|k-1} \,.$$

Setting $x = x_k$, (1) yields

(7) $$\hat{x}_{k|k} = \hat{x}_{k|k-1} + (E\, x_k \epsilon_k')(R_k^{\epsilon})^{-1}(y_k - H_k \hat{x}_{k|k-1})$$

where

$$E\, x_k \epsilon_k' = E\, x_k (\tilde{x}_{k|k-1}' H_k' + v_k')$$

(8) $$= P_{k|k-1} H_k' \triangleq K_k$$

We see that, knowing the predicted estimates $\{\hat{x}_{k|k-1}\}$ and the prediction error variances $\{P_{k|k-1}\}$, we can calculate the filtered estimates $\{\hat{x}_{k|k}\}$ and error variances $\{P_{k|k}\}$. It is not hard to see that

(9) $$P_{k|k} = P_{k|k-1} - K_k (R_k^{\epsilon})^{-1} K_k'$$

Under the assumption (4) therefore, the main thing is to compute the predicted estimates $\{\hat{x}_{k|k-1}\}$ and ea-¹ ›f these can be found from the basic formula (2) as

$$\hat{x}_{k|k-1} = \sum_{0}^{k-1} (E\, x_k \epsilon_j')(E\, \epsilon_j \epsilon_j')^{-1} \epsilon_j \tag{10}$$

Since what we are estimating, namely x_k, changes with k, we cannot expect to relate the estimate $\hat{x}_{k|k-1}$ to the previous estimate $\hat{x}_{k-1|k-2}$ and so on. However if $\{x_k\}$ changes in some known way, or partly known way, rather than arbitrarily, then we may be able to find some relations between successive predicted estimates.

State-Space Signal Models and the Kalman Filter

For example, if we know that

$$x_{k+1} = \Phi_k x_k + \Gamma_k u_k , \quad k \geqslant 0 \tag{11a}$$

$$E\, u_k v_\ell' \equiv 0, \quad E\, u_k u_\ell' = Q_k \delta_{k\ell}, \quad E\, u_k x_0' \equiv 0, \quad E\, x_0 x_0' = \Pi_0 \tag{11b}$$

where the $\{\Phi_k\}$, $\{\Gamma_k\}$, $\{Q_k\}$ and Π_0 are known matrices, and if we add this assumption to the earlier assumption (4), namely that

$$y_k = H_k x_k + v_k , \; E\, v_k v_\ell' = R_k \delta_{k\ell} , \; E\, x_k v_\ell' \equiv 0 \tag{11c}$$

then it is easy to see that

$$\hat{x}_{k+1|k} = \Phi_k \hat{x}_{k|k} + \Gamma_k \hat{u}_{k|k} = \Phi_k \hat{x}_{k|k} \tag{12}$$

since $\hat{u}_{k|k} = 0$. Combining this with (7) gives the recursion for predicted estimates

$$\hat{x}_{k+1|k} = \Phi_k \hat{x}_{k|k-1} + \Phi_k K_k (R_k^\epsilon)^{-1} (y_k - H_k \hat{x}_{k|k-1}) , \quad \hat{x}_{0|-1} = 0 \tag{13}$$

Now, if we define

$$\Sigma_{k|k-1} \triangleq E\, \hat{x}_{k|k-1} \hat{x}_{k|k-1}' ,$$

the fact that the innovations are white yields (cf. Eq. (4.16))

$$\Sigma_{k+1|k} = \Phi_k \Sigma_{k|k-1} \Phi_k' + \Phi_k K_k (R_k^\epsilon)^{-1} K_k' \Phi_k' , \quad \Sigma_{0|-1} = 0 \tag{14}$$

Then combining this with the recursion for the variance matrix of x_k,

$$\Pi_{k+1} = \Phi_k \Pi_k \Phi_k' + \Gamma_k Q_k \Gamma_k'$$

gives

$$P_{k+1|k} = \Pi_{k+1} - \Sigma_{k+1|k}$$

(15) $$= \Phi_k P_{k|k-1} \Phi_k' + \Gamma_k Q_k \Gamma_k' - \Phi_k K_k (R_k^{\epsilon})^{-1} K_k' \Phi_k', \quad P_{0|-1} = \Pi_0 .$$

Equations (13) and (15), along with the relations (6) and (8) for R_k^{ϵ} and K_k, define the celebrated Kalman filter [3]. We restate these equations for convenience.

Predicted Estimates Form of the Kalman Filter

For the state-space model, (11), one-step predicted estimates can be calculated via the recursions

(16a) $$\hat{x}_{k+1|k} = \Phi_k \hat{x}_{k|k-1} + \Phi_k K_k (R_k^{\epsilon})^{-1} (y_k - H_k \hat{x}_{k|k-1}), \hat{x}_{0|-1} = 0$$

where(*)

(16b) $$\epsilon_k = y_k - \hat{y}_{k|k-1}$$

(16c) $$R_k^{\epsilon} = E\, \epsilon_k \epsilon_k' = H_k P_{k|k-1} H_k' + R_k, \quad K_k = P_{k|k-1} H_k'$$

(16d) $$P_{k|k-1} = E\, \tilde{x}_{k|k-1} \tilde{x}_{k|k-1}', \quad \tilde{x}_{k|k-1} = x_k - \hat{x}_{k|k-1}$$

(16e) $$P_{k+1|k} = \Phi_k P_{k|k-1} \Phi_k' + \Gamma_k Q_k \Gamma_k' - \Phi_k K_k (R_k^{\epsilon})^{-1} K_k' \Phi_k', \quad P_{0|-1} = \Pi_0 .$$

The basic equations can be rewritten in several different forms. For example, we can try to get recursions involving only filtered estimates rather than predicted estimates, or forms in which both estimates are mixed. For the latter case, we go from predicted estimates to filtered estimates by the "**measurement-update**" **equations** (7)-(9) and then from filtered to predicted estimates by the "**time-update**" **equations** (12) and the readily-derived formula

(17) $$P_{k+1|k} = \Phi_k P_{k|k} \Phi_k' + \Gamma_k Q_k \Gamma_k' .$$

(*) When $E\, u_k v_l' = C_k \delta_{kl}$ (rather than $\equiv 0$), the only change is that $K_k = P_{k|k-1} H_k' + \Gamma_k C_k$. Show this !

Some Remarks on the Kalman Filter

There are several other rearrangements and extensions and especially numerous other derivations of the Kalman filter equations. A glance at the technical literature of the last 10-15 years will provide numerous examples of such variations. Recently a formalism from scattering theory – Redheffer's star-product – has been found to be very useful in compactly obtaining and organizing old and new results in discrete-time state-space estimation theory [4]-[6]. The Kalman filter has had many successes but it has so dominated the field in the last decade that several other techniques have been unwisely neglected by many "modern" control and system theorists. We cannot pursue this proposition in detail here, but shall content ourselves with brief remarks on some aspects of the Kalman filter and some alternatives to it.

1. One reason for studying different forms of the Kalman filter is that, though all must be theoretically equivalent, in practice they may have different numerical properties. In this regard, certain "square-root" algorithms seem to be the most promising (cf. [7]-[8] and the references therein).

2. It should be stressed that the Kalman filter is obtained as a result of several assumptions, especially (4) and (11), which are by no means necessary for obtaining recursive solutions. The basic recursive formula (1) shows that what is important is having a convenient way of recursively determining the innovations of the observed process. The state-space assumption is one that is very helpful in this regard, but in many problems such models are not readily at hand. In such cases, much effort can often be wasted by first trying to obtain such models so that the Kalman filter can then be applied. However, it is often easier to proceed directly. In this regard, the simple noncausal Wiener filter of Section 2.5 is often overlooked. We mention without comment that the so-called Levinson-Whittle-Wiggins-Robinson algorithm [9]-[11] for stationary processes (cf. the brief discussion in App. II) can often be appropriate, and in fact has been widely used in seismic and geophysical analysis and in speech processing. Generalized Levinson algorithms for nonstationary processes can also be obtained, and can be shown to naturally include the Kalman filter recursions when state-space models are available [11] - [13] (also see App. I).

3. A major feature of the Kalman filter is that it applies to vector observations (y a p-vector) and to time-variant state-space models (11). These problems had been difficult to handle in the earlier given-covariance integral equation approaches and the new assumption of a state-space model and the acceptance (made possible by the development of computers) of an algorithmic

rather than an explicit solution were the keys to the revival of least-squares estimation theory in the late fifties and early sixties. However, armed with the insights gained from the state-space theory, one can go back to the given covariance integral equation approach and obtain equivalent results (cf. ch. VIII).

4. We have noted that the Kalman filter is the same (i.e., requires essentially the same computation) whether or not the observations are scalar or the model parameters are time-variant. However, this very strength is in a sense also a weakness because the Kalman filter cannot exploit the simplifications that can arise (cf. the Wiener theory) if the model parameters are constant and the observed process is scalar and stationary ! In Chapter IX and elsewhere we show how to derive alternative solutions that can specifically exploit this feature. (See also Appendices I and II).

Exercise 5.1.

Suppose $\{y_i\}$ is a stationary process with $E\, y_i y_j = \rho^{|i-j|}$, $0 < \rho < 1$. Show that the innovations can be found as

$$\epsilon_k = y_k - \rho y_{k-1}, \quad k = 1, 2, \ldots .$$

Use this fact to find a simple explicit formula for R^{-1}, where

$$R = [E\, y_i y_j]_{i,j=1}^{N} .$$

Exercise 5.2. Some Kalman Filter Formulas

A number of different identities and formulas can be obtained for the quantities defining the Kalman filter. Most of them are easily obtained by using the projection theorem. Establish the following :

i) $P_{i|i} = P_{i|i-1} - K_i (R_i^{\epsilon})^{-1} H_i P_{i|i-1}$,

where

$$P_{i|i} = E\, \tilde{x}_{i|i} \tilde{x}'_{i|i}, \quad \tilde{x}_{i|i} = x_i - \hat{x}_{i|i}$$

ii) $P_{i|i} = [I - K_i (R_i^{\epsilon})^{-1} H_i] P_{i|i-1} [I - K_i (R_i^{\epsilon})^{-1} H_i]' + K_i (R_i^{\epsilon})^{-1} R_i (R_i^{\epsilon})^{-1} K_i'$

iii) If $R_i > 0$,

$$K_i(R_i^{\epsilon})^{-1} = P_{i|i}H_i'R_i^{-1},$$

$$\hat{x}_{i|i} = P_{i|i}[P_{i|i-1}^{-1}\hat{x}_{i|i-1} + H_i'R_i^{-1}y_i].$$

$$P_{i|i}^{-1} = P_{i|i-1}^{-1} + H_i'R_i^{-1}H_i.$$

These expressions arise in some classical statistical approaches to recursive filtering –the Bayesian approach (see, e.g., [14]) or the least-squares approach (see, e.g., [15]).

The matrix identity

$$[A + BCD]^{-1} = A^{-1} - A^{-1}B[C^{-1} + DA^{-1}B]^{-1}DA^{-1}$$

can be useful in obtaining alternative formulas.

Exercise 5.3. Direct Equations for Filtered Estimates.

Assume that $E\, u_i v_j' \equiv 0.$

Show that we can write

i) $\epsilon_i = y_i - H_i\hat{x}_{i|i-1}$

$$= y_i - H_i\Phi_{i,i-1}\hat{x}_{i-1|i-1} = H_i\Phi_{i,i-1}\tilde{x}_{i-1|i-1} + H_i\Gamma_i u_{i-1} + v_i$$

and

$$R_i^{\epsilon} = H_i\Phi_{i,i-1}P_{i-1|i-1}\Phi_{i,i-1}'H_i' + H_i\Gamma_i Q_{i-1}\Gamma_i'H_i' + R_i$$

where

$$\tilde{x}_{i-1|i-1} \triangleq x_{i-1} - \hat{x}_{i-1|i-1},\quad P_{i|i} \triangleq E\,\tilde{x}_{i|i}\tilde{x}_{i|i}'$$

ii) $\hat{x}_{i+1|i+1} = \Phi_{i+1,i}\hat{x}_{i|i} + K_{i+1}(R_{i+1}^{\epsilon})^{-1}\epsilon_{i+1}$

$$\hat{x}_{0|0} = \Pi_0 H_0'(H_0\Pi_0 H_0' + R_0)^{-1}y_0$$

iii) $P_{i+1|i+1} = \Phi_{i+1,i}P_{i|i}\Phi_{i+1,i}' + \Gamma_{i+1}Q_i\Gamma_{i+1}' - K_{i+1}(R_{i+1}^{\epsilon})^{-1}K_{i+1}',$

$$P_{0|0} = \Pi_0 - \Pi_0 H_0'(H_0\Pi_0 H_0' + R_0)^{-1}H_0\Pi_0$$

iv) $\hat{x}_{i+1|i} = \Phi_{i+1,i}\hat{x}_{i|i}$

v) $P_{i+1|i} = \Phi_{i+1,i}P_{i|i}\Phi'_{i+1,i} + \Gamma_{i+1}Q_i\Gamma'_{i+1}, \quad P_{0|-1} = \Pi_0 .$

NOTE: We see that the **direct** equations for $\hat{x}_{i|i}$ and $P_{i|i}$ are quite complicated. We have given them partly to show how much difference can arise from a slight change in the point of view.

Exercise 5.4. Alternative Model

In Chapter 4, we noted an alternative "u_{i+1}" model for a (discrete-time) process $\{y_i\}$, assumed here for simplicity as time invariant:

$$x_{i+1} = \Phi x_i + \Gamma u_{i+1} \qquad (\text{"}u_{i+1}\text{"-model})$$
$$y_i = H x_i + v_i$$

with

$$E x_0 = 0, \; E u_i = 0, \; E v_i = 0$$

$$E x_0 x_0' = \Pi_0, \; E u_i u_j' = Q\delta_{ij}, \; E v_i v_j' = R\delta_{ij},$$

$$E u_i v_j = C\delta_{ij}, \; E v_i x_0' = 0, \; E u_i x_0' = 0 \text{ for } i \geq 0 .$$

i) Find the filtered estimates of x_i, i.e.,
$\hat{x}_{i|i}$ = function of the previous estimate $\hat{x}_{i-1|i-1}$ and of the innovations $\epsilon_i = y_i - \hat{y}_{i|i-1}$.
(Note that $\hat{y}_{i|i-1} = H\hat{x}_{i|i-1} = H\Phi\hat{x}_{i-1|i-1}$).

ii) Also find the equations for the gains and the error covariance

$$P_{i|i} \triangleq E\tilde{x}_{i|i}\tilde{x}'_{i|i}, \; \tilde{x}_{i|i} \triangleq x_i - \hat{x}_{i|i} .$$

The results for this model are more useful for studies of spectral factorization and system identification [2] [13].

Exercise 5.5.

Prove that the two arrays shown below can be transformed into each other by orthogonal transformations applied to the columns:

$$\begin{bmatrix} R_i^{\frac{1}{2}} & H_iP_i^{\frac{1}{2}} & 0 \\ 0 & \Phi_i P_i^{\frac{1}{2}} & \Gamma_i Q_i^{\frac{1}{2}} \end{bmatrix} = \text{the "pre-array"}, \ C_1$$

$$\begin{bmatrix} (R_i^{\epsilon})^{\frac{1}{2}} & 0 & 0 \\ \overline{K}_i & P_{i+1}^{\frac{1}{2}} & 0 \end{bmatrix} = \text{the "post-array"}, \ C_2$$

$$\overline{K}_i = \Phi_i P_i H_i' (R_i^{\epsilon})^{-\frac{1}{2}}, \quad P_i \triangleq P_{i|i-1} \cdot$$

HINT: Verify that the "squares" of the two arrays are the same, $C_1C_1' = C_2C_2'$. Square-root algorithms have nice numerical properties and are further studied in [7] - [8], (where other references can be found) .

Exercise 5.6. Residuals & Innovations

i) The filtered residuals of

$$y_i = H_ix_i + V_i$$

are defined as

$$\mu_i = y_i - H_i\hat{x}_{i|i} \cdot$$

Show that the $\{\mu_i\}$ form a white-noise sequence with variance matrix

$$R_i^{\mu} = R_i - H_iP_{i|i}H_i'.$$

ii) Show that $\mu_i = R_i(R_i^{\epsilon})^{-1} \epsilon_i$

iii) Show that the random processes

$$\mu_i^j = y_i - H_i\hat{x}_{i|j}, \text{ any } j > i, \text{ have similar properties.}$$

NOTE : We see that many residual white-noise sequences can be obtained from $\{y_i\}$. However the strictly causal 1-step - prediction - error residuals $\{\epsilon_i = y_i - H_i\hat{x}_{i|i-1}\}$ are the most significant for least-squares estimation. This is another reason for giving these residuals a special name, the innovations.

REFERENCES

[1] T. Kailath, "An Innovations Approach to Least-Squares Estimation, Part I : Linear Filtering in Additive White Noise", **IEEE Trans. Automatic Control,** vol. AC-13, pp. 646-655, December 1968.

[2] M. Gevers and T. Kailath, "An Innovations Approach to Least-Squares Estimation, Part VI : Discrete-Time Innovations Representations and Recursive Estimation", **IEEE Trans. Automatic Control,** vol. AC-18, pp. 588-600, December 1973.

[3] R.E. Kalman, "A New Approach to Linear Filtering and Prediction Problems", **J. Basic Eng.,** vol. 82, pp. 34-45, March 1960.

[4] L. Ljung, T. Kailath and B. Friedlander, "Scattering Theory and Linear Least-Squares Estimation, Part I : Continuous-Time Problems", **Proc. IEEE**, vol. 64, No. 1, pp. 131-139, January 1976.

[5] B. Friedlander, T. Kailath and L. Ljung, "Scattering Theory and Linear Least Squares Estimation, Part II : Discrete-Time Problems", **J. of The Franklin Institute,** vol. 301, Nos. 1 & 2, pp. 71-82, January-February 1976.

[6] B. Friedlander, G. Verghese and T. Kailath, "Scattering Theory and Linear Least-Squares Estimation", Part III: The Estimates, Proceedings 1977 **IEEE Decision and Control Conference,** New Orleans, La, Dec. 1977. Also IEEE Trans. Automat. Control, vol. AC-25, pp. 794-802, Aug. 1980.

[7] M. Morf and T. Kailath, "Square-Root Algorithms for Least-Squares Estimation", **IEEE Trans. Automatic Control,** vol. AC-20, No. 4, pp. 487-497, August 1975.

[8] G. Bierman, **Factorization Methods for Discrete Sequential Estimation,** New York : Academic Press, 1977.

[9] N. Levinson, "The Wiener rms (root-mean-square) Error Criterion in Filter Design and Prediction", **J. Math. Phys.,** vol. 25, pp. 261-278, January 1947.

[10] E.A. Robinson, **Multichannel Time-Series Analysis with Digital Computer Programs,** San Francisco, Ca. : Holden-Day 1967.

[11] B. Friedlander, M. Morf, T. Kailath and L. Ljung, "New Inversion Formulas for Matrices Classified in Terms of Their Distance From Toeplitz **Matrices", J. Lin. Alg. and Applns., vol. 27, pp. 31-60, Oct. 1979.**

[12] B. Friedlander, T. Kailath, M. Morf and L. Ljung, "Levinson – and Chandrasekhar – type Equations for a General Discrete-Time Linear Estimation Problem", Proceedings 1976 IEEE Decision and Control Conference, Florida, December 1976. Also IEEE Trans. Automat. Control, pp. 653-659, Aug. 1978.

[13] B. Dickinson, M. Morf and T. Kailath, "Canonical Matrix Fraction and State Space Description for Deterministic and Stochastic Linear Systems", **IEEE Trans. on Automatic Control,** Special Issue on System Identification and Time-Series Analysis, vol. AC-19, No. 6, pp. 656-667, December 1974.

[14] P. Swerling, **Uncertain Dynamic Systems,** Englewood Cliffs : Prentice-Hall, 1973.

[15] F. Schweppe, "Modern State Estimation Methods from the Viewpoint of the Method of Least Squares", **IEEE Trans. Automatic Control,** Vol. AC-16, pp. 707-720, December 1971.

6. CONTINUOUS–TIME KALMAN FILTERS

There are several methods of approaching this problem, of which we shall discuss only two [1], [2]. Since we already have a solution in discrete-time, it is natural to try to use a limiting procedure to get the continuous-time formulas. We shall first show how to do this. However, then we shall show that a solution can be obtained even more directly by using the innovations process.

6.1. Discrete-Time Approximations

There are many methods of obtaining discrete-time approximations to continuous-time problems. Here we shall describe one that has the virtues of being simple and of allowing an easy deduction of continuous-time estimation formulas from their discrete counterparts; it is not the best one from the viewpoint of numerical computation.

The approximation procedure is based on approximating a function m(.) as ($\doteq$ denotes approximate equality)

(1) $$m(t) = \sum_i m_i p(t - i\Delta) + o(\Delta) \doteq \sum_i m_i p(t - i\Delta)$$

where

(2) $$p(t) = \begin{cases} 1/\sqrt{\Delta}, & 0 \leqslant t \leqslant \Delta \\ 0, & \text{elsewhere} \end{cases}$$

and

$$\Delta = \text{an arbitrary (usually small) interval}$$

(3) $$m_i = \int m(t)\, p(t - i\Delta)\, dt / \int p^2(t - i\Delta) dt$$

(4) $$\doteq m(i\Delta)\sqrt{\Delta}, \quad \text{if } m(\cdot) \text{ is continuous}$$

We shall set

$$t_i = i\Delta,$$

so that our approximation for continuous functions is

(5) $$m(t) \doteq \sum_i m(t_i)\sqrt{\Delta} p(t - i\Delta).$$

Then for differentiable functions, we shall have

$$\frac{d}{dt} m(t) = \dot{m}(t) \doteq \sum_i \dot{m}(t_i)\sqrt{\Delta}p(t - i\Delta)$$
$$\doteq \sum_i (m(t_{i+1}) - m(t_i))\, p(t - i\Delta)/\sqrt{\Delta} \tag{6}$$

Also note the product formula

$$m_1(t)m_2(t) \doteq \sum_i m_{1i}m_{2i}p(t - i\,\Delta)/\sqrt{\Delta} \tag{7}$$
$$\doteq \sum_i m_1(t_i)m_2(t_i)\sqrt{\Delta}p(t - i\Delta), \text{for continuous functions} \tag{8}$$

Consider now a random process n(.) with

$$E\ n(t) \equiv 0, \quad E\ n(t)n(s) = R(t,s). \tag{9}$$

Then we have the approximation

$$n(t) \doteq \sum_i n_i p(t - i\,\Delta) \tag{10}$$

where

$$E\ n_i = 0, \quad E\, n_i n_j = \iint R(t, s)\, p(t - i\,\Delta)\, p(s - j\Delta)\, dt\, ds. \tag{11}$$

If the process has continuous paths and R(t,s) is also continuous in t and s, then we can express n(.) in terms of the sample values,

$$n_i \doteq n(t_i)\sqrt{\Delta} \tag{12}$$

so that (as can also be obtained from (11))

$$E\, n_i n_j \doteq \Delta\, R(t_i\,,\, t_j) \tag{13}$$

For white noise u(.) we cannot use the sample-value representation because the paths of u(.) are not continuous. However, note that if

$$E\ u(t)u(s) = Q(t)\delta(t - s) \tag{14}$$

then by (11)

$$E\ u_i\, u_j = Q(t_i)\delta_{ij} \tag{15}$$

This suggests that we can, for uniformity with the other notations, still write

$$u_i \doteq u(t_i)\sqrt{\Delta} \tag{16}$$

provided we agree to define the $\{u(t_i)\}$ as random variables such that

$$E\, u(t_i) = 0, \quad E\, u(t_i)\, u(t_j) = \frac{Q_i}{\Delta}\,\delta_{ij} \tag{17}$$

This is not a very sophisticated approximation of continuous-time white noise, but as we shall see, it enables us to go easily from discrete-time filtering formulas to their continuous-time counterparts. Moreover, we might note that a process of the form

$$u(t) = \sum_i u(t_i)\, p(t - i\Delta - \epsilon)\sqrt{\Delta}, \quad f(\epsilon) = 1/\Delta$$

with the $\{u(t_i)\}$ having the above statistics has been used as a model for thermal noise (cf. Lawson and Uhlenbeck [3, p. 69]).[$f(\epsilon)$ is the probability density function of ϵ].

Finally we note that though we have only discussed scalar functions and scalar processes, the extension to the vector case is obvious.

6.2. Application to the Filtering Problem

$$(18) \qquad \begin{aligned} \dot{x}(t) &= F(t)x(t) + G(t)u(t), \quad t > 0 \\ y(t) &= H(t)x(t) + v(t), \quad x(0) = x_0 \end{aligned}$$

where

$$(19) \qquad \begin{aligned} &E\,x_0 = 0, \quad E\,x_0 x_0' = \Pi_0, \quad E\,u(t)x_0' \equiv 0, \quad E\,v(t)x_0' = 0 \\ &E\,u(t)u'(s) = Q(t)\delta(t-s), \quad E\,v(t)v'(s) = R(t)\delta(t-s) \\ &E\,u(t)v'(s) \equiv C(t)\delta(t-s), \quad R(t) > 0\,. \end{aligned}$$

The problem is to calculate

$$\hat{x}(t) = \text{the 1.1.s.e. of } x(t) \text{ given } \{y(s),\ s < t\}.$$

Using the approximations discussed above we obtain the discrete-time problem described by

$$(20) \qquad \begin{aligned} x(t_{i+1}) &= (I + F(t_i)\Delta)x(t_i) + G(t_i)\Delta u(t_i) \\ y(t_i) &= H(t_i)x(t_i) + v(t_i), \quad x(0) = x_0 \end{aligned}$$

where

$$E\,u(t_i)u'(t_j) = (Q(t_i)/\Delta)\delta_{ij}$$

$$(21) \qquad E\,v(t_i)v'(t_j) = (R(t_i)/\Delta)\delta_{ij}$$

$$E\, u(t_i)v'(t_j) = (C(t_i)/\Delta)\delta_{ij}$$

The predicted estimate based on this model can be written (cf. Eqs. (5.16))

$$\hat{x}(t_{i+1}|t_i) = (I + F(t_i)\Delta)\hat{x}(t_i|t_{i-1}) + K_{g,i}\Delta(y(t_i) - H(t_i)\hat{x}(t_i|t_{i-1})) \qquad (22)$$

where

$$K_{g,i} = [(I + F(t_i)\Delta)P(t_i|t_{i-1})H'(t_i) + G(t_i)C(t_i)][R_i^{\epsilon}]^{-1}\Delta^{-1}$$

$$R_i^{\epsilon} = [H(t_i)P(t_i|t_{i-1})H'(t_i) + R(t_i)/\Delta] \qquad (23)$$

$$P(t_i|t_{i-1}) \triangleq E\,\tilde{x}(t_i|t_{i-1})\tilde{x}'(t_i|t_{i-1}) .$$

Now in the limit as

$$i \to \infty,\ \Delta \to 0,\quad t_i = i\Delta \to t \qquad (24)$$

we see that

$$t_{i-1} \to t,\ \hat{x}(t_i|t_{i-1}) \to \hat{x}(t|t),$$

$$P(t_i|t_{i-1}) \to P(t|t),$$

$$K_{g,i} \to [P(t|t)H'(t) + G(t)C(t)]R^{-1}(t) \triangleq K(t) . \qquad (25)$$

The estimator equation can be written

$$\frac{\hat{x}(t_{i+1}|t_i) - \hat{x}(t_i|t_{i-1})}{\Delta} = F(t_i)\hat{x}(t_i|t_{i-1}) + K_{g,i}(y(t_i) - H(t_i)\hat{x}(t_i|t_{i-1})), \qquad (26)$$

which in the limit becomes the differential equation

$$\dot{\hat{x}}(t|t) = F(t)\hat{x}(t|t) + K(t)[y(t) - H(t)\hat{x}(t|t)] \qquad (27)$$

By using the difference equation (5.15) for $P(t_{i+1} | t_i)$, we see that in the limit $P(t|t)$ obeys the matrix Riccati differential equation

$$\dot{P}(t|t) = F(t)P(t|t) + P(t|t)F'(t) + G(t)Q(t)G'(t)$$

$$- K(t)R(t)K'(t),\ P(0|0) = \Pi_0,\ K(t) = [P(t|t)H'(t) + G(t)C(t)]R^{-1}(t) \qquad (28)$$

Equations (27) and (28) are the celebrated Kalman filter formulas.

For compactness, $\hat{x}(t|t)$ is often just written $\hat{x}(t)$ and $P(t|t)$ as $P(t)$.

Example 6.2.1 : Parameter Estimation

$$\dot{x} = 0, \quad y = x + v, \quad \overline{x}_0 = 0, \quad \Pi_0 = 10$$

The Riccati equation

$$\dot{P} = -P^2, \quad P(0) = 10$$

can be solved to give

$$P(t) = 10/(10t + 1)$$

so that

$$\dot{\hat{x}}(t) = \frac{10}{10t + 1} \upsilon(t), \quad \hat{x}(0) = 0 .$$

Example 6.2.2 : Ornstein-Uhlenbeck Signal Process

$$\dot{x} = -\alpha x + u, \quad Q = 2\alpha, \quad C = 0, \quad R = 1$$
$$y = x + v, \quad \Pi_0 = 1$$

Also let

$$\gamma = \alpha \sqrt{1 + 2/\alpha}$$

The Riccati equation

$$\dot{P} = -2\alpha P - P^2 + 2\alpha, \quad P(0) = 1$$

has a unique positive solution

$$P(t) = \frac{2\alpha}{\gamma + \alpha} \frac{1 + \left(\dfrac{\gamma - \alpha}{\gamma + \alpha}\right) e^{-2\gamma t}}{1 - \left(\dfrac{\gamma - \alpha}{\gamma + \alpha}\right)^2 e^{-2\gamma t}} \underset{t \to \infty}{\longrightarrow} \gamma - \alpha$$

Note that in the steady state, the filter is

$$\dot{\hat{x}} = -\alpha \hat{x} + (\gamma - \alpha)(y - \hat{x}),$$

which corresponds to a transfer function

$$H(s) = \alpha\left(\sqrt{1 + \frac{2}{\alpha}} - 1\right) \frac{1}{s + \alpha\sqrt{1 + \frac{2}{\alpha}}}$$

that agrees with that of the Wiener filter for this problem (cf. Example 3.3.3).

Exercise 6.2.1 : Noise-Free Systems (No Process Noise)

If

$$G \equiv 0 \text{ and } \Pi_0 > 0$$

show that

i) $$P(t) = \Phi(t, t_0)[\Pi_0^{-1} + \mathscr{I}(t, t_0)]^{-1}\Phi'(t, t_0)$$

where $\Phi(t,t_0)$ is the state-transition matrix and

$$\mathscr{I}(t, t_0) = \int_{t_0}^{t} \Phi'(s, t_0)H'(s)H(s)\Phi(s, t_0)\, ds$$

$\mathscr{I}(t, t_0)$ is sometimes called the Fisher information matrix for reasons related to the result in (iv) below.

ii) $$P(t_0 \mid T) = [\Pi_0^{-1} + \mathscr{I}(T, t_0)]^{-1}$$

iii) If we reinitialize the filtering problem with $\hat{x}(t_0 | T)$ and $P(t_0 | T)$, then

$$P(t_0 \mid 2T) = [\Pi_0^{-1} + 2\mathscr{I}(T,t_0)]^{-1}$$

iv) If $\Pi_0 = \infty$, i.e., no prior information, then

$$P(t_0 \mid T) = \mathscr{I}^{-1}(T, t_0)\text{, a classical result.}$$

These results can be proved by verification.

Exercise 6.2.2.

Let

$$y_t = x_t + v_t, \qquad t \geqslant 0$$

where v_t is a white noise process,

$$\dot{x}_t = \alpha x_t\,, \qquad t \geqslant 0$$

and x_0 is a Gaussian random variable with zero mean and variance P_0.

a) Show that

$$P_{t/t} \triangleq E[x_t - \hat{x}_{t/t})^2]$$

$$= \frac{e^{2\alpha t} P_0}{1 + \frac{P_0}{2\alpha}(e^{2\alpha t} - 1)}$$

and examine the limiting behavior as $t \to \infty$ for both $\alpha > 0$ and $\alpha < 0$.

b) Compute

$$P_{s/t} = E[(x_s - \hat{x}_{s/t})^2]$$

and comment on the behavior as $t \to \infty$ for both $\alpha > 0$ and $\alpha < 0$.

c) Compare results of Parts a) and b).

Exercise 6.2.3.

Given a system

$$\dot{\mathcal{E}}(t) = a\mathcal{E}(t), \quad \mathcal{E}(0) = \mathcal{E}_0$$

$$y(t) = \sum_{K=1}^{N} h_k \mathcal{E}^k(t) + v(t)$$

where $\mathcal{E}(t)$ is scalar and $\mathcal{E}^k(t)$ is the k-th power of $\mathcal{E}$, show how to choose N states so as to obtain an equivalent N-state linear system,

$$\dot{\mathbf{x}}(t) = \mathbf{F}\,\mathbf{x}(t), \quad \mathbf{x}(0) = \mathbf{x}_0, \quad \overline{x_0 x_0'} = \Pi_0$$

$$y(t) = \mathbf{H}\,\mathbf{x}(t) + v(t)$$

Show that the error covariance P(t) is given by $P(t) = e^{Ft}\{\Pi_0^{-1} + A\}^{-1} e^{F't}$, where A is a matrix with (i, j)-th element

$$\frac{h_i h_j}{a(i+j)} \left(e^{a(i+j)t} - 1\right).$$

Exercise 6.2.4.

Suppose

$$y(t) = z(t) + v(t), \quad t_0 = -\infty$$

$$\ddot{z}(t) + 3\dot{z}(t) + 2z(t) = u(t)$$

where u(.) and v(.) are independent white-noise processes with intensities σ_1^2 and σ_2^2 respectively.

i) Find the variance of z(t) and $\dot{z}(t)$?

ii) Construct a constant coefficient linear system with y(.) as input that will give $\hat{z}(t\,|t)$ as output.

Exercise 6.2.5.

A body leaves the initial position x_0 at time $t = 0$ with velocity v_0 and acceleration a_0. We assume that the position thereafter is governed by the equation

$$x(t) = x_0 + v_0 t + a_0 (t^2/2)$$

and that x_0, v_0, and a_0 are independent random variables with means $\bar{x}_0$, $\bar{v}_0$, $\bar{a}_0$ and variances σ_x^2, σ_v^2 and σ_a^2 respectively. Also assume that the position $x(.)$ is measured continuously according to the equation

$$y(t) = x(t) + v(t)$$

where $v(.)$ is a unit intensity white noise independent of x_0, v_0, and a_0.

Calculate $\hat{x}(s|t)$ for all $0 \leq s, t \leq T$, i.e., determine filtered, predicted and smoothed estimates. Do not solve the equations.

6.3. Relation to Deterministic Observers

At about the same time as the Kalman equations were being developed, a method of estimating states in deterministic systems was proposed by J. Bertram and R. Bass and was later developed by D. Luenberger into a theory of "observers" (cf. Astrom [4]). We shall show that the Kalman filter can be obtained in a natural way as an extension of this theory.

Given

$$\dot{x} = Fx + Gu_0$$

$$y = Hx + v_0$$

with known F, G, H, u_0, v_0 and y, but unknown $x(0)$, the deterministic observer has the form

$$\dot{\hat{x}}_d = F\hat{x}_d + Gu_0 + K_d[y - H\hat{x}_d - v_0], \quad \hat{x}_d(0) = 0$$

where in the constant-parameter case, the constant K_d is chosen to make the poles of the system have specified locations, thus enabling control of the rate of decay of the error. In the general time-variant noisy case, we proceed as follows: Suppose we actually have

$$y = Hx + v_0 + v, \quad \dot{x} = Fx + Gu_0 + Gu$$

where u and v are zero-mean white noises of intensities Q and I and $x(0)$ is random with mean m and variance Π_0. Then we should choose K_d so as to minimize the variance of the error $\tilde{x}_d$. We have

$$\dot{\tilde{x}}_d = (F - K_d H)\tilde{x}_d + Gu - K_d v, \; \tilde{x}_d(0) = x_0$$

The error mean obeys the differential equation

$$\frac{d}{dt} E\,\tilde{x}_d = (F - K_d H) E\,\tilde{x}_d, \; E\,\tilde{x}_d(0) = m$$

and the error variance obeys (cf. Eq. (4.12))

$$\dot{P}_d = (F - K_d H)P_d + P_d(F - K_d H)' + GQG' + K_d K_d', \; P_d(0) = \Pi_0$$

The problem is to choose K_d so as to minimize P_d in the sense that $P_{d_1} \geqslant P_{d_2}$ if $P_{d_1} - P_{d_2}$ is nonnegative definite. This minimization can be done directly. However, from our previous results on the Kalman filter we know that the optimum value for K_d must be

$$K_{d,0} = PH'$$

where

$$\dot{P} = FP + PF' + GQG' - K_{d,0}K_{d,0}', \; P(0) = \Pi_0$$

To verify that $P_d \geqslant P$, we set

$$\mathcal{E} = P_d - P$$

and note that

$$\begin{aligned}
\dot{\mathcal{E}} &= F\mathcal{E} + \mathcal{E}F' + K_d K_d' - K_d H P_d - P_d H' K_d' + K_{d,0}K_{d,0}' \\
&= (F - K_d H)\mathcal{E} + \mathcal{E}(F - K_d H)' - K_d HP - PH' K_d + K_d K_d' + PH'HP \\
&= (F - K_d H)\mathcal{E} + \mathcal{E}(F - K_d H)' + (K_d - PH')(K_d - PH')', \; \mathcal{E}(0) = 0 .
\end{aligned}$$

Let

$$\psi(t, s) = \text{the state-transition matrix of } F - K_d H .$$

Then

$$\mathcal{E}(t) = \int_0^t \psi(t, \tau) A(\tau) \psi'(t, \tau) \, d\tau$$

where

$$A(\tau) = [K_d(\tau) - P(\tau)H'(\tau)][K_d(\tau) - P(\tau)H'(\tau)]' .$$

However $A(\tau)$ is clearly nonnegative definite, so that

$$\mathcal{E}(t) \geqslant 0$$

with equality only when

$$A(\tau) = 0, \text{ i.e., } K_d(\tau) = P(\tau)H'(\tau).$$

6.4. The Innovations Process in Continuous-Time

The process

$$\nu(t) = y(t) - H(t)\hat{x}(t \mid t) \tag{1}$$

$$= y(t) - \hat{y}(t \mid t-) \tag{2}$$

is the continuous-time limit of the discrete-time innovations process $\{\epsilon_i\}$. Even in continuous-time, it continues to have the innovations interpretation as the new information in the observation at time t, viz, y(t), after all information, $\hat{y}(t \mid t-)$, in past observations $\{ y(s), s \leqslant t- \}$ has been removed. In view of this intuitive significance it is not surprising that (as in discrete-time), $\nu(.)$ is a white-noise process. However, this continuous-time innovations process $\nu(.)$ has an interesting property that makes the continuous-time Kalman filter have a simpler form than its discrete-time counterpart. We note that

$$\epsilon(t_i) = y(t_i) - H(t_i)\hat{x}(t_i \mid t_{i-1}) = H(t_i)\tilde{x}(t_i \mid t_{i-1}) + v(t_i) \tag{3}$$

and

$$E\, \epsilon(t_i)\epsilon'(t_j)' = \left[H(t_i)P(t_i \mid t_{i-1})H'(t_i) + \frac{R(t_i)}{\Delta} \right] \delta_{ij}$$

$$= \frac{R(t_i) + \Delta H(t_i)P(t_i \mid t_{i-1})H'(t_i)}{\Delta}\, \delta_{ij}\,. \tag{4}$$

Now comparing this with (17) and (14) shows that in the continuous-time limit as $\Delta \to 0$

$$\{\epsilon(t_i)\} \to \text{a white-noise process, } \nu(t) \tag{5}$$

with

$$E\,\nu(t)\nu'(s) = R(t)\,\delta(t-s). \tag{6}$$

Note that since R(t) is assumed positive-definite, we can always normalize it to be the identity matrix. This will be assumed henceforth. But the interesting thing is that in continuous-time, the intensity of the innovations process $\nu(.)$ is the same as that of the additive white-noise v(.) in the original observed process y(.).

This fact can also be seen by a direct calculation, which we shall only give in a heuristic form here. [Several rigorous proofs can be given and some can be found in [5] - [8], see also Exercise 6.4.1.] Thus note that

$$\nu(t) = y(t) - \hat{y}(t\,|\,t\text{-}) = y(t) - \hat{z}(t) \tag{7}$$

$$= \tilde{z}(t) + v(t).$$

Therefore we can write

$$E\,\nu(t)\nu'(s) = E\,v(t)v'(s) + A(t,s) \tag{8}$$

where

$$A(t,s) \triangleq E[v(t)\tilde{z}'(s) + \tilde{z}(t)\nu'(s)]. \tag{9}$$

Now by the assumption that future v(.) are uncorrelated with past signal and noise and also by use of the projection theorem, we see that

$$A(t,s) = 0 + 0 = 0, \quad \text{for } t > s \tag{10}$$

Since we can rewrite A(t,s) as

$$A(t,s) = E[\tilde{z}(t)v'(s) + \nu(t)\tilde{z}'(s)]$$

we see by a similar argument that

$$A(t,s) = 0, \quad \text{for } t < s. \tag{11}$$

What about $t = s$? Here we have to appeal to the fact that if the signal process z(.) is "smoother" than the noise (e.g., if z(.) has finite expected energy), then A(t,s) will be a continuous function and therefore from (10)-(11) we could conclude that

$$A(t,s) = 0 \quad \text{for all } t,s \tag{12}$$

and therefore

$$E\,\nu(t)\nu'(s) = E\,v(t)v'(s) = I\,\delta(t-s). \tag{13}$$

The smoothness requirements on z(.) are actually quite weak, e.g., if z(.) and v(.) are uncorrelated it suffices to assume that z(.) has finite expected energy. But actually even weaker constraints suffice (cf. [5]- [8]), but to discuss and prove these would take too much machinery here. Therefore we shall be content with the above heuristic but simple proof (see also Exercise 6.4.1.).

As in discrete-time, we may expect that the limiting continuous-time innovations would contain the same information as the original observed process, so that we can go back from $\nu(.)$ to y(.) by a causal operation. Of course this fact can also be directly proved in continuous-time following the lines suggested in Exercise 6.4.2.

There are many fascinating aspects of the continuous-time innovations process. For example, we might mention that if

$$y(t) = z(t) + v(t),$$

with

$$v(t) = \text{white Gaussian noise}$$
$$z(t) = \text{a smooth but not necessarily Gaussian process}$$

then

$$\nu(t) = y(t) - \hat{z}(t),$$
$$\hat{z}(t) = \text{least-squares estimate (not necessarily linear) of } z(t)$$
$$\text{given} \quad \{y(s),\ s \leqslant t-\}$$

is not only a white noise process with the same intensity as v(.) but $\nu(.)$ is also Gaussian [5] [7] [8].

This is rather remarkable since both z(.) and y(.) (and hence $\hat{z}(.)$) can be highly non-Gaussian, e.g., z(.) could be Poisson.

To really understand the above result will require us to go into martingale theory and a closer study of concepts like sigma-fields, etc. Actually, such a study is of value even in the linear problem, for example, in trying to understand the relation between the two white processes v(.) and $\nu(.) = v(.) + \tilde{z}(.)$. A tutorial exposition of such matters and some applications of the resulting insights are given in [9]. Here we shall content ourselves with showing how the innovations process $\nu(.)$ can be used to give a very simple derivation of the Kalman filter. The simplicity of this derivation has been helpful in tackling several variants of the Kalman filter problem.

Exercise 6.4.1. Covariance of the Innovations Process

Let

$$\hat{z}(t) = \int_0^t h(t, s) y(s) ds$$

where the optimum causal filter H(.,.) obeys the Wiener-Hopf-type equation

$$h(t, s) + \int_0^t h(t, \tau) K(\tau, s) d\tau = K(t, s), \quad 0 \leqslant s \leqslant t \leqslant T.$$

Now compute E $\nu(t)\nu(s)$ using only the facts that

$$y(t) = z(t) + v(t), \quad E\, v(t) z(s) \equiv 0, \quad t > s$$
$$E\, y(t) y(s) = \delta(t - s) + K(t, s).$$

Hint: $\displaystyle\int_{-\infty}^{t} \delta(t - \tau) d\tau + \int_t^{\infty} \delta(t - \tau) d\tau = 1.$

Exercise 6.4.2. Equivalence of Innovations and Observations [2, App. II], [6], [20].

To show that y(.) can be recovered from ν(.), note that

$$\nu(t) = \int_0^t [\delta(t-s) - h(t,s)]\, y(s)\, ds$$

can be solved for y(.) by the "Neumann" series in symbolic notation

$$y = (I + h + h^2 + \ldots)\nu ,$$

or explicitly

$$y(t) = \nu(t) + \int_0^t h(t,s)\nu(s)\, ds + \int_0^t h(t,\tau) d\tau \int_0^\tau h(\tau,s)\nu(s) ds + \ldots$$

It is known that this expansion will be valid if

$$\int_0^T \int_0^T h^2(t,s)\, dt\, ds < \infty .$$

Show that this condition on h(.) can be ensured by the assumption

$$E \int_0^T z^2(t) dt < \infty .$$

Exercise 6.4.3.

In the Kalman filter notation let

$$\nu(t) = H(t)\tilde{x}(t) + v(t) ,$$

where the error covariance $P(t) = E\, x(t)\tilde{x}'(t)$ is known to obey the Riccati equation

$$\dot{P}(t) = F(t)P(t) + P(t)F'(t) - P(t)H'(t)H(t)P(t) + G(t)Q(t)G'(t),$$

$$P(0) = \Pi_0 .$$

Prove by direct calculation using this formula for P(.) that

$$E\,\nu(t)\nu'(s) = I\,\delta(t-s) .$$

Remark: The point is that this result holds in great generality, whether or not we have a state model. And in fact, it has a somewhat simpler general proof than the above. Moreover, as we shall show in Section 6.5, the general result can be used to easily derive the Riccati equation for the Kalman filter.

Exercise 6.4.4.

Consider the process

$$y_t = \mu + v_t$$

where v_t is a white noise and μ is an independent Gaussian random variable with mean m and variance σ^2. Define υ_t by

$$\upsilon_t = y_t - \hat{\mu}|t$$

Show that this innovations process is equivalent to the y_t process by constructing y_t from $\{\upsilon_\sigma, \sigma \leqslant t\}$.

6.5. Innovations Derivation of the Continuous-Time Kalman Filter

At this point it will be useful to see how the innovations process $\nu(.)$ can be exploited to directly give the Kalman filter. To simplify certain arguments, we shall here assume that

$$C(t) = 0, \quad \text{i.e.,} \quad u(\cdot) \perp v(\cdot) \tag{14}$$

The general case can of course also be treated with a little more effort (cf. [2], [6]).

Step.1. Introduce the Innovations

Define the innovations $\nu(.)$ via

$$\nu(t) = y(t) - \hat{z}(t) = y(t) - H(t)\hat{x}(t) = H(t)\tilde{x}(t) + v(t) \tag{15}$$

Step.2. Estimates from the Innovations

Let

$$\hat{x}(t) = \int_0^t g(t,s)\nu(s)\, ds . \tag{16}$$

By the projection theorem, g(t,s) must be such that

$$x(t) - \hat{x}(t) \perp \nu(\tau), \quad \tau < t, \tag{17}$$

Therefore

$$E\, x(t)\nu'(\tau) = \int_0^t g(t,s)\, \overline{\nu(s)\nu'(\tau)}\, ds, \qquad \tau < t$$

$$= g(t,\tau), \qquad \tau < t \tag{18*}$$

(*) Bars are used for convenience to denote expectation.

so that

(19) $$\hat{x}(t) = \int_0^t \overline{x(t)\nu'(s)}\nu(s)\ ds\,,\quad \hat{x}(0) = 0\,.$$

Now we use the fact that x(t) obeys the state equation
$$\dot{x}(t) = F(t)x(t) + G(t)u(t)$$
which suggests(*) that we differentiate the above expression for $\hat{x}(t)$ to get

$$\dot{\hat{x}}(t) = \overline{x(t)\nu'(t)}\nu(t) + \int_0^t \overline{\dot{x}(t)\nu'(s)}\nu(s)\ ds$$

(20) $$= K(t)\nu(t) + F(t)\hat{x}(t) + G(t)\int_0^t \underbrace{\overline{u(t)\nu'(s)}}_{=\,0,\quad s<t}\nu(s)\ ds\,.$$

Also

(21) $$K(t) = \overline{x(t)\nu'(t)} = \overline{x(t)\tilde{x}'(t)}H'(t) + \underbrace{\overline{x(t)v'(t)}}_{=\,0} = P(t)H'(t)$$

since

(22) $$P(t) = E\ \tilde{x}(t)\ \tilde{x}'(t)\ .$$

These are the equations of the Kalman filter, except for the specification of P(.).

Step 3. Variance Calculations

If

(23) $$\Sigma(t) = \overline{\hat{x}(t)\hat{x}'(t)}\,,\quad \Pi(t) = \overline{x(t)x'(t)}$$

then since

(24) $$x(t) = \hat{x}(t) + \tilde{x}(t) \quad\text{and}\quad \tilde{x}(t) \perp \hat{x}(t)\,,$$

it follows that

(25) $$\Pi(t) = P(t) + \Sigma(t)\,.$$

Now since both x(.) and $\hat{x}$(.) obey state-equations driven by white-noise, their variances obey the differential equations (cf. Section IV),

(26) $$\dot{\Pi} = F\Pi + \Pi F' + GQG',\ \Pi(0) = \Pi_0$$

(27) $$\dot{\Sigma} = F\Sigma + \Sigma F' + KK',\ \Sigma(0) = 0$$

(*) The rigorous form of this argument appears in [6].

so that

$$\dot{P} = \dot{\Pi} - \dot{\Sigma} = FP + PF' - KK' + GQG', \; P(0) = \Pi_0 .$$

This is the matrix Riccati equation, which is usually derived somewhat less directly (see Exercise 6.5.1.).

The relations (25)-(27) are very simple but are quite important and will be exploited later, e.g., in Section 8. There have been numerous derivations of the Kalman filter, but the above approach (the rigorous form of which appears at [6]) appears to be the simplest and most far reaching. A justification of this statement is that numerous variants of the Kalman filter can also be very easily derived by the innovations approach. Some early examples were given in [2], [10] but several others have appeared since then, e.g., we name only the recent ones [11]- [12]

Related Problems

There are numerous special aspects of the state-space estimation problem that can and have been studied: for example, different methods of solving the Riccati equation, variations of the basic filtering algorithm to accommodate special needs or constraints, recursive formulas for smoothing, etc. There is a vast literature on these issues, which we shall not really enter into here, except for the few results on smoothing noted in the exercises and the topics of the remaining chapters. We may, however, draw special attention to Exercise 6.5.4. which provides a starting point for a so-called scattering approach to state-space estimation. Our experience is that this formulation provides the best way of studying and organizing the numerous special problems of continuous- and discrete-time state-space estimation (see [13] - [19]).

Exercise 6.5.1. Error Equation and Riccati Equation

Find a differential equation for the error $\tilde{x}(.)$ and hence obtain a differential equation for the variance $P(.)$.

Exercise 6.5.2. A Basic Smoothing Formula [10]

Given observations $\{y(t) = H(t)x(t) + v(t)\}$ over $t_0 \leqslant t \leqslant t_f$ show that we can write

$$\hat{x}(t \mid t_f) = \hat{x}(t \mid t) + \int_t^{t_f} P(t, s)H'(s)\nu(s)ds$$

where

$$\nu(t) = y(t) - H(t)\hat{x}(t \mid t)$$

$$P(t, s) = E\,\tilde{x}(t \mid t)\tilde{x}(s \mid s) \quad .$$

Show also that the smoothed error variance

$$P(t \mid t_f) = E\,\tilde{x}(t \mid t_f)\tilde{x}'(t \mid t_f)$$

$$= P(t \mid t) - \int_t^{t_f} P(t, s)H'(s)H(s)P'(t, s)ds \, .$$

These formulas show that smoothed estimates are completely determined by filtered estimates.

Exercise 6.5.3. Some Smoothing Formulas for State Models

The result in the previous exercise does not assume any state model for x(.). If we have the usual model show that

i) $P(t, s) = P(t)\Phi_K'(s, t)$

where Φ_K is the state-transition matrix of the closed-loop filter, F – KH.

ii) $\hat{x}(t \mid t_f) = \hat{x}(t \mid t) + P(t)\lambda(t)$

where

$$\dot{\lambda}(t) = -[F(t) - K(t)H(t)]'\lambda(t) - H'(t)\nu(t)\, ; \; \lambda(t_f) = 0.$$

iii) when $C(.) \equiv 0$, show that we can write

$$\frac{d}{dt}\hat{x}(t \mid t_f) = F(t)\hat{x}(t \mid t_f) + G(t)Q(t)G'(t)P^{-1}(t)\cdot[\hat{x}(t \mid t_f) - \hat{x}(t \mid t)] \, , \; t \leqslant t_f \, .$$

This formula is due to Rauch, Tung and Striebel and is just one of several smoothing formulas that can be obtained from the basic formula of Exercise 6.5.3. [10]. Insights from scattering theory have led to several new families of smoothing formulas [18]

Exercise 6.5.4. The Hamiltonian Equations

When $C(.) \equiv 0$, show that we can write

$$\begin{bmatrix} \dot{\hat{x}}(t|t_f) \\ -\dot{\lambda}(t) \end{bmatrix} = \begin{bmatrix} F(t) & G(t)Q(t)G'(t) \\ -H'(t)H(t) & F'(t) \end{bmatrix} \begin{bmatrix} \hat{x}(t|t_f) \\ \lambda(t) \end{bmatrix} + \begin{bmatrix} 0 \\ H'(t)y(t) \end{bmatrix}.$$

These equations have the form of the "Hamiltonian equations" often encountered in the calculus of variations. They allow us to set up a so called "Redheffer" scattering model (see [13], [14]) for the state-space estimation problem and thereby provide a very efficient way of studying a large variety of special state-space estimation problems (see [15]- [19]).

Exercise 6.5.5. A Simple "Singular Noise" Problem

Find a differential equation for $\hat{x}(t|t)$ when

$$F = \begin{bmatrix} 0 & 1 \\ 0 & -2 \end{bmatrix}, \quad G = \begin{bmatrix} 1 \\ 0 \end{bmatrix}$$

$$H = \begin{bmatrix} 1 & 0 \\ 1 & 2 \end{bmatrix}, \quad J = \begin{bmatrix} 1 \\ 0 \end{bmatrix}, \quad \Pi_0 = \begin{bmatrix} 2 & 1 \\ 1 & 2 \end{bmatrix}$$

HINT: Note that $y_2(.)$ is a noise free measurement and can therefore be differentiated.

REFERENCES

[1] R.E. Kalman, "New Methods of Wiener Filtering Theory", in **Proc. 1st Symp. Engineering Applications of Random Function Theory and Probability**, J.L. Bogdanoff and F. Kozin, Eds., New York: Wiley, 1963, pp. 270-388; also, RIAS, Baltimore, Md., Tech. Rep. 61-1, 1961.

[1a] R.E. Kalman and R.S. Bucy, "New Results in Linear Filtering and Prediction Theory", **Trans. ASME**, Ser. D, **J. Basic Eng.**, vol. 83, pp. 95-107, December 1961.

[2] T. Kailath, "An Innovations Approach to Least-Squares Estimation, Pt. I: Linear Filtering in Additive White Noise", **IEEE Trans. on Automatic Control**, vol. AC-13, pp. 646-655, December 1968.

[3] J. Lawson and G.E. Uhlenbeck, **Threshold Signals**, MIT Radiation Laboratory Series, vol. 24, McGraw-Hill Book Co., New York, 1948.

[4] K.J. Åstrom, **Introduction to Stochastic Control Theory**, Academic Press, New York, 1970.

[5] T. Kailath, "A General Likelihood-Ratio Formula for Random Signals in Gaussian Noise," **IEEE Trans. on Information Theory**, vol. IT-15, pp. 350-361, May 1969.

[6] T. Kailath, " A Note on Least-Squares Estimation by the Innovations Method," **J. SIAM Control**, vol. 10 pp. 477-486, August 1972.

[7] T. Kailath, "Some Extensions of the Innovations Theorem," **Bell Syst. Tech. J.**, vol. 50, pp. 1487-1494, April 1971.

[8] P.A. Meyer, "Sur un problème de filtration", Séminaire de probabilités, Pt. VII, **Lecture Notes in Mathematics**, vol. 321, Springer, New York, pp. 223-247, 1973.

[9] A. Segall, "Stochastic processes in estimation theory", IEEE Trans. Inform. Theory, vol. IT-22, pp. 275-286, 1976.

[10] T. Kailath and P. Frost, "An Innovations Approach to Least-Squares Estimation, Pt. II: Linear Smoothing in Additive White Noise," **IEEE Trans. on Automatic Control**, vol. AC-13, pp. 655-660, December 1968.

[11] S. Fujishige, "Minimum-Variance Estimation for a Linear Continuous-Discrete System with Noisy State-Integral Observation", **IEEE Trans. on Automatic Control**, vol. AC-20, pp.139-140, February 1975.

[12] J. Mishra and V.S. Rajamani, "Least-Squares State Estimation in Time-Delay Systems with Colored Observation Noise: An Innovations Approach", **IEEE Trans. on Automatic Control**, vol. AC-20, pp. 140-142, February 1975.

[13] R. Redheffer, "On the Relation of Transmission-Line Theory to Scattering and Transfer", **J. Math. Phys.**, vol. 41, pp. 1-41, 1962.

[14] R. Redheffer, Difference Equations and Functional Equations in Transmission-Line Theory", Ch. 12 in **Modern Mathematics for the Engineer**, 2nd series (E.F. Beckenbach, ed.), McGraw-Hill Book Co., New York, 1961.

[15] G. Verghese, B. Friedlander and T. Kailath, "Scattering Theory and Linear Least-Squares Estimation" Part III: The Estimates, Proceedings 1977 IEEE Decision and Control Conference, New Orleans, Dec. 1977. Also IEEE Trans. Automat. Contr., vol. AC-25, pp. 794-802, Aug. 1980.

[16] L. Ljung, T. Kailath and B. Friedlander, "Scattering Theory and Linear Squares Estimation" Part I: Continuous-Time Problems, **Proc. IEEE**, vol. 64, pp. 131-139, January 1976.

[17] B. Friedlander, T. Kailath and L. Ljung, "Scattering Theory and Linear Least Squares Estimation" Part II: Discrete Time Problems, Journal of The Franklin Institute, vol. 301, Nos. 1&2, pp. 71-82, January-February 1976.

[18] L. Ljung and T. Kailath, "A Unified Approach to Smoothing Formulas", Automatica, vol. 12, pp. 147-157, March 1976.

[19] B.C. Lévy, D.A. Castanon, G.C. Verghese and A.S. Villsky, A Scattering Framework for Decentralized Estimation Problems, Proc. 1981 IEEE Decision and Control Conference, San Diego, Ca, Dec. 1981.

[20] M.H.A. Davis, Linear Estimation and Stochastic Control, Halsted Press-Wiley, New York, 1977.

7. RELATIONS TO WIENER FILTERS

The Kalman-Bucy filters for state-space models of (asymptotically) stationary processes must tend as $t \to \infty$ (or equivalently $t_0 \to -\infty$) to the Wiener filters of Section 3. We shall study some aspects of this particular limiting case here. In particular we shall derive the so-called Yovits-Jackson closed form expression for the mean-square error of Wiener filters from our results. This formula has attracted some attention in modulation theory (Viterbi [1], Van Trees [2]). More general results on the convergence of the Kalman filter for not necessarily asymptotically stationary processes can be found in [3].

7.1. State-Space Models for Stationary Processes

We shall study the behavior of the Kalman-Bucy filter for the constant-parameter model (for scalar y(.), for convenience),

(1a) $$\dot{x}(t) = F\, x(t) + G\, u(t), \quad x(t_0) = x_0$$

(1b) $$z(t) = H\, x(t), \quad y(t) = z(t) + v(t)$$

where

(1c) $$E\, x_0 x_0' = \Pi_0, \quad E\, u(t) x_0' \equiv 0, \quad t > t_0$$

(1d) $$E\, u(t) u'(s) = Q\, \delta(t-s), \quad E\, v(t) v(s) = \delta(t-s)\,.$$

For easy comparison with the Wiener filtering formulas of Sec. 3, we shall assume that the signal process z(.) is scalar and that it is uncorrelated with the observation noise v(.), or equivalently that

(2) $$E\, u(t) v(s) \equiv 0.$$

To obtain a stationary signal process in the limit as $t_0 \to \infty$, we must also assume that

(3) $$F \text{ is a stability matrix}$$

or equivalently that

$$\mathrm{Re}\, \lambda_i(F) < 0, \quad i = 1, \ldots, n$$

where the $\{\lambda_i(F)\}$ are the eigenvalues of F.

With these assumptions, we note that

(4) $$R_x(t,s) = E\, x(t) x'(s) = \begin{cases} e^{F(t-s)} \Pi(s)\,, & t \geq s \\ \Pi(t) e^{F'(s-t)}\,, & t \leq s \end{cases}$$

where

$$\Pi(t) \triangleq E\ x(t)x'(t)$$

$$= e^{F(t-t_0)}\ \Pi_0\ e^{F'(t-t_0)} + \int_{t_0}^{t} e^{F(t-\tau)} GQG'\ e^{F'(t-\tau)}\ d\tau \tag{5}$$

or equivalently

$$\dot{\Pi}(t) = F\Pi(t) + \Pi(t)F' + GQG', \quad \Pi(0) = \Pi_0 . \tag{6}$$

Also if

$$z(t) = H\ x(t), \tag{7}$$

then

$$R_z(t,s) = H\ R_x(t,s)H' . \tag{8}$$

Now if we assume that

$$t_0 \rightarrow -\infty$$

we may expect that

$$\Pi(t) \rightarrow \overline{\Pi}, \quad \text{a constant matrix} \tag{9}$$

and that $\overline{\Pi}$ obeys the algebraic matrix equation

$$0 = F\overline{\Pi} + \overline{\Pi}F' + GQG' . \tag{10}$$

The fact that F is a stability matrix ensures that this algebraic equation has a unique nonnegative definite solution (lemma of Lyapunov). [Controllability of the pair (F, $GQ^{1/2}$) will ensure positive definiteness [4]].

Note also that as $t_0 \rightarrow -\infty$, z(.) tends to a stationary process with covariance function

$$R_z(t, s) = H\, e^{F(t-s)}\overline{\Pi}H' , \quad t \geqslant s \tag{11}$$

and power-spectral density function

$$S_z(j\omega) = H(j\omega I - F)^{-1}\overline{\Pi}H' + H\overline{\Pi}(-j\omega I - F')^{-1}H' \tag{12}$$

$$= H(j\omega I - F)^{-1} GQG'(-j\omega I - F')^{-1}H' \tag{13}$$

Note however that if we set

$$\Pi_0 = \overline{\Pi} \tag{11a}$$

then $\dot{\Pi} = 0$ for all t and hence

$$\Pi(t) = \overline{\Pi}, \text{ for all } t \geqslant t_0 \tag{14b}$$

That is, the system falls instantly into "steady-state".

7.2. Steady State Kalman Filter for Stationary Processes

The Kalman filter for the constant-parameter model (1) (whether it is in steady-state or not, i.e., whether or not $\Pi_0 = \overline{\Pi}$) is

$$\dot{\hat{x}}(t) = F\ \hat{x}(t) + K(t)[y(t) - H\ \hat{x}(t)], \quad \hat{x}(0) = 0$$

where

$$K(t) = P(t)H'(t)$$

$$\dot{P}(t) = FP(t) + P(t)F' - K(t)K'(t) + GQG',$$

$$= (F - K(t)H)\,P(t) + P(t)(F - K(t)H)' + K(t)K'(t) + GQG',\ P(0) = \Pi_0$$

The matrix P(.) is bounded above because

$$P(t) \leqslant \Pi(t)$$

and $\Pi(t)$ is well behaved when F is a stability matrix.

When

(15) $$\Pi_0 = \overline{\Pi}$$

then it is easy to prove that (cf. Equations (11) and (18) of Section 9)

(16) $$\dot{P}(t) \leqslant 0 \quad \text{all finite } t\,.$$

Therefore we must have

(17) $$P(t) \searrow \overline{P} \text{ as } t \to \infty\,.$$

Moreover it is clear that $\dot{P}(t)$ also converges (obviously to zero) and therefore $\overline{P}$ must obey

(18) $$0 = F\overline{P} + \overline{P}F' - \overline{P}H'H\overline{P} + GQG'.$$

Actually, the convergence (though not necessarily monotonic) of P(t) to $\overline{P}$ can be proved without the assumption $\Pi_0 = \overline{\Pi}$; it is true for arbitrary initial choices $\Pi_0 \geqslant 0$. The convergence also holds even if F is not a stability matrix, provided certain detectability and stabilizability assumptions are imposed (cf. the recent paper [3], which appears to subsume most previous results on this topic).

Therefore, the steady-state Kalman filter is

(19a) $$\dot{\hat{x}}(t) = F\ \hat{x}(t) + \overline{K}(y - H\hat{x}),\ \overline{K} = \overline{P}H'$$

(19b) $$\hat{z}(t) = H\ \hat{x}(t)$$

which has the transfer function from y(.) to $\hat{z}(.)$,

$$\mathcal{H}_K(s) = H(sI - F + \bar{K}H)^{-1}\bar{K}. \tag{20}$$

The Wiener filter for this problem has transfer function (cf. Example 3.3.4)

$$\mathcal{H}_w(s) = 1 - [S_y^+(s)]^{-1} \tag{21}$$

where $S_y^+(s)$ is the canonical spectral factorization of the spectral density of the observed process (cf. (13))

$$\begin{aligned} S_y(s) &= 1 + S_z(s) \\ &= 1 + H(sI - F)^{-1} GQG'(-sI - F')^{-1} H'. \end{aligned} \tag{22}$$

We shall show that

$$S_y^+(s) = 1 + H(sI - F)^{-1}\bar{K}. \tag{23}$$

This follows from using the innovations property to write the Kalman filter equations in the form

$$\dot{\hat{x}}(t) = F\,\hat{x}(t) + \bar{K}\,\nu(t), \quad E\nu(t)\nu'(s) = \delta(t - s) \tag{24a}$$

$$y(t) = H\,\hat{x}(t) + \nu(t). \tag{24b}$$

This form is clearly causal and causally invertible and the $j\omega$-axis is the region of convergence; therefore, by well known facts in Fourier transform theory, the above system is causal and minimum phase, i.e., all its poles and zeros are in the LHP. The transfer function of the above model yields the canonical factor (23).

From the representation (24) for the steady-state Kalman filter we can write the transfer function from y(.) to $\hat{z}(.)$ as the product of the transfer functions from ν(.) to $\hat{z}(.)$ and from y(.) to ν(.), i.e.,

$$\begin{aligned} \mathcal{H}_K(s) &= H(sI - F)^{-1}\bar{K}\cdot[S_y^+(s)]^{-1} \\ &= [S_y^+(s) - 1][S_y^+(s)]^{-1} \\ &= 1 - [S_y^+(s)]^{-1} \\ &= \mathcal{H}_w(s). \end{aligned} \tag{25}$$

Even more directly, note from (24b) that

$$\hat{z}(.) = y(.) - \nu(.) \tag{26a}$$

or

$$\mathcal{H}_K(s) \triangleq \frac{\hat{Z}(s)}{Y(s)} = 1 - \frac{1}{S_y^+(s)} = \mathcal{H}_w(s) \tag{26b}$$

Another method of proving the equality (25) is to use the matrix identity

$$[1 + H(sI - F)^{-1}\bar{K}]^{-1} = 1 - H(sI - F + \bar{K}H)^{-1}\bar{K}. \tag{27}$$

We may note that from (22) and (23) we have the formula

$$\begin{aligned} &1 + H(sI - F)^{-1}GQG'(-sI - F')^{-1}H' \\ &= [1 + H(sI - F)^{-1}\bar{K}][1 + \bar{K}'(-sI - F')^{-1}H'] \end{aligned} \tag{28}$$

which is a special case of an identity, first discovered in stability theory and optimal control theory (Popov [5], Kalman [6]). There are numerous further connections of these results to spectral factorization, network synthesis, control theory (see e.g., [7] - [10], which we shall not pursue here.

7.3. Spectral Factorization and the Steady-State Riccati Equation

It is interesting to pursue further the connections between the canonical spectral factor $S_y^+(s)$ and the solution of the steady-state or algebraic Riccati equation (18).

We begin with a constant parameter model

$$\begin{aligned} \dot{x}(t) &= F\ x(t) + G\ u(t) \\ y(t) &= H\ x(t) + v(t), \quad E\ u(t)v(s) \equiv 0, \end{aligned}$$

assumed to be in the steady-state so that we can write the power-spectral density function of the scalar process y(.) as

$$S_y(s) = 1 + H(sI - F)^{-1}GQG'(-sI - F')^{-1}H'.$$

Now as stated earlier, many models can be associated with a given spectral density function.

It will be convenient to choose a model that is in what is called "observer" canonical form, [4]

$$H = [1\ \ 0\ldots.0],\ \ G' = [\beta_1 \ldots\ldots. \beta_n] \tag{29a}$$

and

$$F = \begin{bmatrix} -\alpha_1 & & 1 & & \bigcirc \\ \cdot & & & 1 & \\ \cdot & & \bigcirc & & \ddots \\ \cdot & & & & \cdot\, 1 \\ -\alpha_n & 0 & 0 & \dots\dots & 0 \end{bmatrix} \tag{29b}$$

The transfer function is directly related to the $\{\alpha_i, \beta_i\}$,

$$H(sI - F)^{-1}G = \beta(s)/\alpha(s) \tag{29c}$$

where

$$\alpha(s) = \det(sI - F) = s^n + \alpha_1 s^{n-1} + \dots + \alpha_n \tag{29d}$$

and

$$\beta(s) = \beta_1 s^{n-1} + \dots\dots + \beta_n\ . \tag{29e}$$

We note also that in observer form

$$H(sI - F)^{-1} = [s^{n-1}\ s^{n-2}\ \dots\ 1]\ \frac{1}{\alpha(s)}\,. \tag{29f}$$

Now the power spectral density of $y(.)$ can be written

$$S_y(s) = 1 + \frac{\beta(s)\beta(-s)}{\alpha(s)\alpha(-s)}\,. \tag{30}$$

In the notation of Sec. 3, the canonical factor is

$$S_y^+(s) = \left[\frac{\alpha(s)\alpha(-s) + \beta(s)\beta(-s)}{\alpha(s)\alpha(-s)}\right]^+ = \frac{[\alpha(s)\alpha(-s) + \beta(s)\beta(-s)]^+}{\alpha(s)} \tag{31}$$

$$= \gamma(s)/\alpha(s)\,,\ \text{say} \tag{32}$$

where $\gamma(s)$ is a polynomial of the form

$$\gamma(s) = s^n + \gamma_1 s^{n-1} + \dots \gamma_n\ . \tag{33}$$

However in terms of the steady-state solution P of the Riccati equation we have (cf. (23))

$$S_y^+(s) = 1 + H(sI - F)^{-1}\bar{P}H'\ .$$

Comparing these two expressions, we get

(34) $$H(sI - F)^{-1}\bar{P}H' = \frac{\gamma(s) - \alpha(s)}{\alpha(s)} .$$

However by the properties of the observer canonical form (cf. (29f)), we see that we can obtain the relation

(35a) $$\bar{P}_{11} s^{n-1} + \ldots + \bar{P}_{n1} = \sum_1^n (\gamma_i - \alpha_i) s^{n-i}$$

or

(35b) $$\bar{P}_{i1} = \gamma_i - \alpha_i, \quad i = 1, \ldots, n .$$

Therefore the first column of $\bar{P}$ is completely determined by knowledge of the canonical factorization. Note that in observer canonical form only the first column of $\bar{P}$ is required to calculate the gain function $\bar{K} = \bar{P}H'$, so that as expected, the Kalman filter is completely determined by the canonical factorization.

Note that if we do not use the observer canonical form, we will have

$$(sI - F)^{-1} = \frac{\varphi(s)}{\alpha(s)}, \quad \text{say}$$

and

$$H\varphi(s) = \gamma(s) - \alpha(s) .$$

From these equations we can still relate the n spectral factorization quantities $\{\gamma_i - \alpha_i\}$ to the $n(n+1)/2$ elements of $\bar{P}$. The observer canonical forms shows how to ignore the $[n(n+1)/2] - n$ "redundant" components of $\bar{P}$ [4].

Mean–Square-Error

The mean-square-error, under the assumption of uncorrelated signal and noise, is

(36) $$\text{m.s.e.} = H\bar{P}H' = H\bar{K}.$$

With the observable canonical form, this is just

$$\text{m.s.e.} = \bar{P}_{11} = \gamma_1 - \alpha_1$$

(37) $$= \sum_1^n (\text{roots of } \gamma(s)) - \sum_1^n (\text{roots of } \alpha(s)) .$$

It can be shown by direct evaluation of the integral that

$$\sum_{1}^{n}[(\text{roots of } \alpha(s)) - (\text{roots of } \gamma(s))] = \int_{-\infty}^{\infty} \ln[1 + S_z(\omega)] \frac{d\omega}{2\pi}, \quad (38)$$

thus giving the formula of Yovits and Jackson that was noted in Eq. (3.9).

The last formula has been used in the analysis of modulation systems (see, e.g., Viterbi [1], Van Trees [2]). However, the other forms may sometimes be more convenient for numerical calculation [10]. Note that, as expected, the m.s.e. can be computed without first having to know the optimum filter.

Alternative Derivation

J. Snyders has pointed out (personal communication, 1973) that it is not necessary to use the observer canonical form. Instead one may do the following.

Let

$$I = \int_{-\infty}^{\infty} \ln(1 + S_z(\omega)) \frac{d\omega}{2\pi}$$

and note that

$$I = \lim_{s \to \infty} \int_{-\infty}^{\infty} \ln(1 + S_z(\omega)) \frac{s}{s - i\omega} \frac{d\omega}{2\pi} .$$

Then using (28) and closing the contour on the right gives

$$I = \lim_{s \to \infty} s \ln[1 + H(sI - F)^{-1}\bar{K}]$$

$$= H\bar{K} = H\bar{P}H' = \text{the m.s.e.}$$

REFERENCES

[1] A.J. Viterbi, **Principles of Coherent Communication,** McGraw-Hill Book Co., New York.

[2] H.L. Van Trees, **Detection, Estimation, and Modulation Theory, Pt. I,** J. Wiley & Sons, New York, 1968. **Detection, Estimation and Modulation Theory, Pt. II: Nonlinear Modulation Theory,** J. Wiley & Sons, New York 1971.

[3] T. Kailath and L. Ljung, "Asymptotic Behaviour of Constant-Coefficient Riccati Differential Equations", **IEEE Trans. on Automatic Control,** vol. AC-21, pp. 385-388, 1976.

[4] T. Kailath, **Linear Systems,** Prentice-Hall, New Jersey, 1980.

[5] V.M. Popov, "Hyperstability and Optimality of Automatic Systems with Several Control Functions", **Rev. Roum. Sci. Tech.,** vol. 9, pp. 629-690, 1964.

[6] R.E. Kalman, "When is a Linear Control System Optimal ? ", **Trans. ASME,** Ser. D., **J. Basic Eng.,** vol. 86, pp. 51-60, June 1964.

[7] T. Kailath, "A View of Three Decades of Linear Filtering Theory", **IEEE Transactions on Information Theory,** vol. IT-20, pp. 145-181, March 1974.

[8] B.D.O. Anderson and S. Vongpanitlerd, **Network Analysis and Synthesis – A Modern Systems Theory Approach,** Prentice-Hall, Inc., New Jersey, 1973.

[9] V.M. Popov, **Hyperstability of Control Systems,** Springer-Verlag, New York, 1973. (Original Rumanian edition, Bucharest, 1966).

[10] P. Van Dooren, "A generalized eigenvalue approach for solving Riccati equations", SIAM J. Scient. and Statist. Computing, vol. 2, pp. 121-135, June 1981.

8. RECURSIVE WIENER FILTERS

We now return to the problem of finding $\hat{z}(t)$ given $y(t) = z(t) + v(t)$, when only the covariance functions $E\ z(t)z'(s)$, $E\ z(t)v'(s)$, $E\ v(t)z'(s)$ and $E\ v(t)v'(s)$ are known.

One approach (Kalman [1]; Faurre [2], [3]) is to construct a lumped model for the signal process $z(.)$ and then to apply the Kalman filtering algorithm. However the problem of finding a model for $z(.)$, even if a unique model is not necessary, is quite difficult.

Another approach (Kailath and Geesey [4], [5]), concentrates on the problem of determining a model for the **observed** process $y(.)$, a particular unique model that yields y as the response of a causal and causally invertible filter to a white-noise input. One motivation for seeking this **innovations representation** (IR) is that we know it is fundamental to Wiener filtering; in fact, the optimum Wiener filter is immediately determined by the IR via the formula

$$\mathcal{H}_{opt}(s) = 1 - \frac{1}{S_y^+(s)}. \tag{1}$$

Also, in fact, we know by the innovations theorem that if

$$y(t) = z(t) + v(t), \text{ then } \hat{z}(t) = y(t) - \nu(t). \tag{2}$$

Therefore determination of the IR of the observations y is the key to the problem.

To find the IR we shall proceed as follows: We shall find that when a model is known for y, the IR can be immediately obtained by a rearrangement of the Kalman filter formulas. Now it is known that the impulse response of the IR of a process is uniquely determined by the covariance function of the process. Therefore it must be possible to re-express the IR obtained from a known model in terms of the covariance function.

Thus suppose we have a model for $y(.)$ of the form

$$\dot{x} = Fx + Gu, \quad y = Hx + v \tag{3}$$

where $u(.)$, $v(.)$, x_0 are defined as in the general Kalman problem (cf. (19)). Then the Kalman filter is

$$\begin{aligned} \dot{\hat{x}} &= F\hat{x} + K(y - H\hat{x}), \quad \hat{x}(0) = 0 \\ K &= PH' + GC, \quad \dot{P} = FP + PF' - KK' + GQG', \quad P(0) = \Pi_0 \end{aligned} \tag{4}$$

However, by rewriting these equations as

$$\dot{\hat{x}} = F\hat{x} + K\nu, \quad \hat{x}(0) = 0 \qquad (5)$$
$$y = H\hat{x} + \nu$$

we see that we have a causal and causally invertible representation of y as the output of a linear system driven by a white noise ν. Thus the IR is determined in terms of the model parameters. However, we know that we must be able to rewrite it in terms of the covariance function of y(.), which is (cf. Section IV)

$$R_y(t,s) = E\,y(t)y'(s) = I\,\delta(t-s) + K(t,s) \qquad (6)$$

where

$$K(t,s) = M(t)\Phi(t,s)N(s)1(t-s) + N'(t)\Phi'(s,t)M'(s)1(s-t)\,, \qquad (6a)$$

$$M(t) = H(t), \quad N(t) = \Pi(t)M'(t) + G(t)C(t)\,, \qquad (6b)$$

$$\Pi(t) = E\,x(t)x'(t), \quad \dot{\Pi}(t) = F\Pi + \Pi F' + GQG', \quad \Pi(0) = \Pi_0\,, \qquad (6c)$$

$$\frac{d\Phi(t,s)}{dt} = F(t)\Phi(t,s), \qquad \Phi(s,s) = I\,. \qquad (6d)$$

To re-express the above form of the IR in terms of M, Φ, N, we first define

$$\Sigma(t) = E\,\hat{x}(t)\hat{x}'(t) = \text{the variance of the states of the IR}\,, \qquad (7)$$

and note that since

$$x(t) = \hat{x}(t) + \tilde{x}(t), \quad \text{where} \quad \tilde{x}(t) \perp \hat{x}(t) \qquad (8)$$

we have

$$\Pi(t) = \Sigma(t) + P(t)\,. \qquad (9)$$

Therefore, we can write

$$K(t) = PH' + GC = \Pi H' + GC - \Sigma H'$$
$$= N - \Sigma M'\,. \qquad (10)$$

Moreover from (9), (4) and (6) we see that

$$\dot{\Sigma} = \dot{\Pi} - \dot{P} = F\Sigma + \Sigma F' + KK'\,, \quad \Sigma(0) = 0 \qquad (11)$$

so that $\Sigma(.)$ is described by a Riccati equation specified completely by the covariance parameters $\{F,N,M\}$. The existence of a solution to this nonlinear Riccati equation (11) is ensured by the inequality

$$0 \leqslant \Sigma(t) \leqslant \Pi(t) \tag{12}$$

We see therefore that the IR has now been completely expressed in terms of the covariance parameters,

$$\dot{\hat{x}} = F\hat{x} + K\nu, \quad y = M\hat{x} + \nu, \quad K = N - \Sigma M' \tag{13a}$$

$$\dot{\Sigma} = F\Sigma + \Sigma F' - [N - \Sigma M'][N - \Sigma M']', \; \Sigma(0) = 0. \tag{13b}$$

Also the estimate of z(.) is recursively determined as

$$\hat{z} = y - \nu = M\hat{x}. \tag{14}$$

That is, we have a recursive solution for $\hat{z}$ directly expressed in terms of the covariance parameters. In other words, we have now obtained a recursive solution of the generalized Wiener-Hopf problems posed in Section IV. Note especially that there is no need to first determine a state-space model for z(.) and then apply a Kalman filter – not only would this be wasteful but the problem of determining a model is not a simple one.

In fact, it turns out [10a] [10b] that trying to find a model for z(.) will essentially require the solution of another Riccati equation like (11). Then another Riccati equation like (4) would have to be solved to obtain the Kalman filter for the derived model for z(.). However the important fact shown above is that by finding a special innovations model for y(.) (and not z(.)), no more work will be needed to obtain the estimate $\hat{z}$(.).

The point is that we now have two possible state-space formulations of the estimation problem – we may be given a state-space description for the signal process z(.) and the noise process v(.) (i.e., we may be given F,G,H,Q,Π_0) or we may be given only a state-space description (i.e. F,M,N) of the covariance functions of z(.) and v(.). In the first case, we should use the Kalman filter (4) ; in the second, we should use the equations (13)–(14).

Recursive Solution of Integral Equations

In Section IV, we posed the estimation problem in terms of a generalized Wiener-Hopf equation, but even for separable kernels we noted that the known solution methods were quite cumbersome. Since we have now found a recursive (and presumably computationally effective) way of finding the estimate, we might expect that there is a recursive way of finding the solution of the generalized Wiener-Hopf equation. The answer of course is yes and in fact, the solution h(t, τ) is the impulse response of the system

$$\dot{\hat{x}} = F\hat{x} + K(y - M\hat{x}), \quad \hat{x}(0) = 0$$

$$\hat{z} = M\hat{x},$$

where K, M are as in (12)-(13). For further discussion of the integral equation we refer the reader to [6]-[9] and to Exercises 8.1. and 8.2.

However, a final comment is important. In the above arguments for the given-covariance problem, we used the fact that there was some (not necessarily known) state-space model generating the given (separable) covariance. This assumption is not necessary and all the above results can be proved without it (see [10a, Appendix III]). However, these arguments are purely mathematical and lack the important insights (the IR is unique and therefore the obvious IR obtained by rewriting the Kalman filter must be reexpressible in terms of the covariance) provided by our presentation above. As an example of the advantages of having such insights we may note the solution of a smoothing problem given in [4], whereas it had seemed via other approaches [11] that no smoothing solutions could be found. For a somewhat simpler illustration, we leave the following exercise to the reader.

Let w(t) be a process related to y(.) by

$$E\, w(t)y'(s) = M_w(t)\Phi(t,s)N(s), \quad t \geqslant s.$$

Prove that

$$\hat{w}(t) = M_w(t)\hat{x}(t)$$

where $\hat{x}(.)$ is as in (13).

Exercise 8.1. The Siegert-Krein-Bellman Identity [12]

Let

$$R(t,s) = \delta(t-s) + K(t,s), \quad 0 \leqslant t, s \leqslant T$$

be a strictly positive-definite covariance function on [0,T] x [0,T] and assume that K(t,s) is differentiable in both variables. Then, there is a unique differentiable solution to the Fredholm integral equation of the second kind

$$H(t, s; T) + \int_0^T H(t, \tau; T)K(\tau, s)\, d\tau = K(t, s), \qquad 0 \leqslant t, s \leqslant T$$

i) Prove that the Fredholm resolvent H(t,s;T) obeys the following (Siegert-Krein-Bellman) identity

$$\frac{\partial H(t, s; T)}{\partial T} = -H(t, T; T)H(T, s; T).$$

ii) Define h(t,s) as the unique solution (differentiable for $t > s$) of the Wiener-Hopf equation

$$h(t, s) + \int_0^t h(t, \tau)K(\tau, s)\, d\tau = K(t, s), \qquad 0 \leqslant s \leqslant t$$

$$h(t, s) = 0, \qquad s < t.$$

Show that

$$\frac{\partial H(t, s; T)}{\partial T} = -h^*(t, T)h(T, s)$$

where

$$h^*(t, s) = h(s, t).$$

iii) Show that the integrated forms of the above relations are

$$\begin{aligned} H(t, s; T) &= H(t, s; t) + H(s, t; s) - \int_0^T H(t, \tau; \tau)H(\tau, s; \tau)\, d\tau \\ &= h(t, s) + h^*(t, s) - \int_0^T h^*(t, \tau)h(\tau, s)\, d\tau. \end{aligned}$$

Exercise 8.2. The Sobolev Identity [13]

Suppose in the previous problem that K(t,s) is a function only of $|t - s|$.

i) Show that

$$H(t, s; T) = H(T - t, T - s; T).$$

ii) Now show that we can write

$$\frac{\partial H(t, s; T)}{\partial t} + \frac{\partial H(t, s; T)}{\partial s} = A(T; t)A(T; s) - B(T; t)B(T; s)$$

where

$$A(T; t) = H(t, T; T) = H(T, t; T) = h(T, t)$$

$$B(T\,;t) = H(t,0\,;T) = H(0,t\,;T) = A(T\,;T-t)\,.$$

iii) Show that

$$H(t,s\,;T) = B(T\,;t-s) + B(T\,;s-t) + \int_0^{\min(t,s)} [A(T\,;t-r)A(T\,;s-r) - \\ - B(T\,;t-r)B(T\,;s-r)]\,dr\,.$$

The significance of these formulas is that they express the resolvent in terms of functions of one variable, which can be implemented by time-invariant filters, thus permitting the use of fast convolution and fast Fourier transform techniques.

The $A(.\,;.)$ and $B(.\,;.)$ functions can also be efficiently computed via certain differential equations first obtained by M.G. Krein. These formulas and several important generalizations thereof are noted in the survey paper of Appendix I and its references.

REFERENCES

[1] R.E. Kalman, "Linear Stochastic Filtering-Reappraisal and Outlook **Proc. Symp. System Theory,** (J. Fox, ed.), Polytechnic Institute of Brooklyn Press, Brooklyn, New York, 1965.

[2] P. Faurre, "**Representation of Stochastic Processes**", Ph.D. Dissertation, Dept. of Electrical Engineering, Stanford University, Stanford, Ca., 1967.

[3] P. Faurre, "Identification par minimisation d'une représentation Markovienne de processus aléatoire", **Lecture Notes in Mathematics,** vol. 132, pp. 85-106, Springer-Verlag, New York, 1970.

[4] T. Kailath, and R. Geesey, "An Innovations Approach to Least-Squares Estimation, Pt. IV: Recursive Estimation Given Lumped Covariance Functions", **IEEE Trans. on Automatic Control,** vol. AC-16, pp. 720-727, December 1971.

[5] T. Kailath, "A View of Three Decades of Linear Filtering Theory", **IEEE Trans. on Information Theory,** vol. IT-20, pp. 145-181, March 1974.

[6] T. Kailath, "Fredholm Resolvents, Wiener-Hopf Equations and Riccati Differential Equations", **IEEE Trans. on Information Theory,** vol. IT-15, pp. 665-672, November 1969.

[7] B.D.O. Anderson and T. Kailath, "The Choice of Signal Process Models in Kalman-Bucy Filtering", **J. Math. Anal. Appl.,** vol. 35, pp. 659-668, September 1971.

[8] T. Kailath, R. Geesey and H. Weinert, "Some Relations Between RKHS Norms, and Fredholm Equations, and Innovations Representations", **IEEE Trans. on Information Theory,** vol. IT-18, pp. 341-348, May 1972.

[9] A. Schumitzky, "On the Equivalence Between Matrix Riccati Equations and Fredholm Resolvents", **J. Comp. Syst. Sci.,** vol. 2, pp. 76-87, June 1968.

[10a] T. Kailath and R. Geesey, "An Innovations Approach to Least-Squares Estimation, Pt. V: Innovations Representations and Recursive Estimation in Colored Noise", **IEEE Trans. on Automatic Control,** vol. AC-18, pp. 435-453, October 1973.

[10b] J.B. Moore and B.D.O. Anderson, "Spectral Factorization of Time-Varying Covariance Functions : The Singular Case", **Mathematical System Theory,** vol. 4, pp. 10-23, 1970.

[11] B.D.O. Anderson and J.B. Moore, "The Kalman-Bucy Filter as a True Time-Varying Wiener Filter", **IEEE Trans. Syst., Man, Cybernetics,** vol. SMC-1, pp. 119-128, 1971.

[12a] M.G. Krein, "On Integral Equations Leading to Differential Equations of Second Order", **Dokl. Akad. Nauk SSSR,** 97, pp. 21-24, 1954.

[12b] T. Kailath, "Application of a Resolvent Identity to a Linear Smoothing Problem", **SIAM J. Contr.,** vol. 7, pp. 68-74, 1969.

[13] V.V. Sobolev, A Treatise on Radiative Transfer, D. Van Nostrand, Princeton, N.J., 1963, (Russian edition 1956).

9. FAST ALGORITHMS FOR CONSTANT PARAMETER MODELS

The widespread use of the Kalman filter in many applications of "modern" control theory led to a great activity in studying the non-linear Riccati equation, which underlies the Kalman filter. Therefore not much attention was paid to the important Wiener-Hopf equation, though its connections to Riccati equations and Fredholm equations were being established, as briefly noted at the end of the last section. Then in [1], preprints of which had circulated earlier, attention was drawn to a different method of solving a certain class of Wiener-Hopf equations. More specifically it was noted that the solution of the equation

$$h(t) + \int_0^\infty h(\tau)K(t-\tau)\, d\tau = K(t), \qquad 0 < t < \infty \tag{1}$$

where K(.) has the form

$$K(t-s) = \int_0^1 \exp - (\alpha|t-s|)\, w(\alpha)\, d\alpha \tag{2}$$

could be obtained in terms of two auxiliary functions obeying the simultaneous nonlinear partial differential equations

$$\frac{\partial X(t,\alpha)}{\partial t} = -Y(t,\alpha)\int_0^1 Y(t,\beta)w(\beta)\, d\beta \tag{3a}$$

$$\frac{\partial Y(t,\alpha)}{\partial t} = -\alpha Y(t,\alpha) - X(t,\alpha)\int_0^1 Y(t,\beta)w(\beta)\, d\beta \tag{3b}$$

$$X(0,\alpha) = 1 = Y(0,\alpha), \quad 0 \leqslant \alpha \leqslant 1. \tag{3c}$$

This was essentially the solution proposed by the astrophysicist S. Chandrasekhar in 1947 [2]. The X and Y functions had been introduced even earlier by Ambartzumian [3], though the differential equations were first given by Chandrasekhar [2] and were used by him and others to obtain numerical solutions by the obvious discretization technique (see, e.g., [4, p. 79]). The stimulus of a seminar by J.L. Casti on the paper [1] led us to explore the possibility of similar results for processes with state-space models with constant coefficients. The first results can be found in [5] - [6], and have since then been much further developed and applied (see [7] - [16] and Appendix I), because they provide algorithms that potentially have several computational advantages over the Riccati based Kalman filter.

To give the flavor of the results, let us consider the problem of state estimation for the model

(4a) $$\dot{x}(t) = F\ x(t) + G\ u(t), \quad x(0) = 0$$

(4b) $$y(t) = H\ x(t) + v(t)$$

where v(.) and u(.) are uncorrelated white-noise processes with

(4c) $$E\ u(t)u'(s) = I\ \delta(t-s), \quad E\ v(t)v'(s) = I\ \delta(t-s)$$

Note that the intensity Q of the plant noise u(.) has been absorbed into the input matrix G. F, G, H, Q are all assumed to be time-invariant, though, of course, this does not mean that the processes x(.) and y(.) are necessarily stationary, as assumed in [1]. Stationarity will be obtained if F is stable and all transients have been allowed to die out, a condition that can also be simulated in finite time by a proper choice of initial condition. This situation will be studied below (cf. (17)), but first let us examine the model (4) in which we note the **assumption that the initial state is known to be zero,** so that the initial variance is 0,

(4d) $$\Pi_o = 0$$

In this case, we shall show that the estimate $\hat{x}(.)$ can be found via the equations

(5) $$\dot{\hat{x}}(t) = F\,\hat{x}(t) + K(t)\,(y(t) - H\,\hat{x}(t)), \quad \hat{x}(0) = 0$$

where K(.) is determined not via the Riccati equation but by the **Chandrasekhar-type** equations

(6a) $$\dot{K}(t) = L(t)L'(t)H', \quad K(0) = 0$$

(6b) $$\dot{L}(t) = [F - K(t)H]L(t), \quad L(0) = G$$

Since K(.) is n x p and L(.) is n x m, we have n(m + p) coupled nonlinear differential equations to determine the gain function, as compared to n(n + 1)/2 to determine the solution P(.) of the Riccati equation. When $n \gg m$, $n \gg p$, this can be a substantial saving in computation.

The formulas (6) can be established in several ways, but we shall give the proof first used [6] to obtain them. This begins with the formulas of the Kalman filter,

(7) $$K(t) = P(t)H'$$

(8) $$\dot{P}(t) = FP(t) + P(t)F' - P(t)H'HP(t) + GG', \quad P(0) = 0$$

Because of the constancy of F, G, H, we can write

(9) $$\dot{K}(t) = \dot{P}(t)H', \quad K(0) = 0$$

$$\begin{aligned}\ddot{P}(t) &= F\dot{P}(t) + \dot{P}(t)F' - \dot{P}(t)H'K(t) - K(t)H'H\dot{P}(t) \\ &= (F - K(t)H)\dot{P}(t) + \dot{P}(t)(F - K(t)H)'\end{aligned} \tag{10}$$

Temporarily regarding K(t) in (10) as a known function, (10) can be regarded as a homogeneous linear differential equations in $\dot{P}(.)$ with solution

$$\dot{P}(t) = \Phi(t,0)\dot{P}(0)\Phi'(t,0) \tag{11}$$

$$= \Phi(t,0)GG'\Phi'(t,0) \tag{12}$$

where $\Phi(t,0)$ in a state-transition matrix defined by

$$\frac{d}{dt}\Phi(t,0) = (F - K(t)H)\Phi(t,0), \quad \Phi(0,0) = I. \tag{13}$$

If we now define

$$L(t) = \Phi(t,0)G \tag{14}$$

we see that

$$\dot{L}(t) = (F - K(t)H)L(t), \quad L(0) = G \tag{15}$$

Now K(.) is not known but from (9), we have

$$\dot{K}(t) = \dot{P}(t)H' = L(t)L'(t)H' \tag{16}$$

which together with (15) forms a closed set of differential equations for K(.). In fact, these are the differential equations (6a)-(6b) as we set out to prove.

Note that the special initial condition P(0) = 0 only enters in giving $\dot{P}(0) = GG'$ and hence $L(0) = G$. We could have any arbitrary value for P(0) but then $\dot{P}(0)$ might not have low rank and there may be no reduction in the number of equations over the Riccati solution [6]. But there are often problems where one has some freedom to choose P(0) and this can be exploited to use a choice that gives computational benefits. This question is discussed in some detail in [14], but here we may note one more special case.

Stationary Processes

If F is stable, then the (Lyapunov) equation

$$F\overline{\Pi} + \overline{\Pi}F' + GG' = 0 \tag{17}$$

is well known to have a unique nonnegative definite solution $\overline{\Pi}$. If $P(0) = \Pi(0) = \overline{\Pi}$, then it can be seen from the formulas of Section IV that x(.) and y(.) will be stationary processes and

$$(18) \qquad \dot{P}(0) = -\bar{\Pi}H'H\bar{\Pi}.$$

Then it is easy to see from (7) and (11) that the appropriate equations will be

$$(19a) \qquad \dot{K}(t) = -L(t)L'(t)H', \quad K(0) = \bar{\Pi}H'$$

$$(19b) \qquad \dot{L}(t) = (F - K(t)H)L(t), \quad L(0) = \bar{\Pi}H'$$

which are now 2np in number. The equations (19a) can be obtained from the Chandrasekhar partial differential equations (3) by assuming $-F = \text{diag}\{\alpha_i\}$, $w(\alpha) = \Sigma\, \alpha_i\, \delta\,(\alpha - \alpha_i)$. Hence the equations (19) and (6) were said to be of Chandrasekhar-type [5], [6].

There is more to say about such equations, but we shall close by noting that a key step in the above proof is the judicious use of the identity (11). This identity was proved in a different way by Bucy (cf. [17, p. 69]) but it was apparently first noted (actually in a more general form) by Bellman, Kalaba and Wing [18]. They label this and certain related identities as Stokes relations, since the basic idea comes from some pioneering investigations of Stokes on the scattering of light [19]. These early results have been developed into a polished theory of reflection and scattering by several authors, especially R. Redheffer and W.T. Reid. As noted earlier in Section 6.5, we have recently found that this theory led us to several new results in estimation theory, including a deeper understanding of the Janus-like relationship between the Riccati-type equations and the Chandrasekhar-type equations.

However we shall not enter into such explorations here, but refer to [10] - [16] and Appendices I and II. The whole now vast topic of fast estimation algorithms essentially grew from this seed.

Exercise 9.1. Chandrasekhar Equations for General Initial Conditions [6]

Let

α = rank $\dot{P}(0)$ = rank $[F\Pi_0 + \Pi_0 F' + GQG' - \Pi_0 H'H\Pi_0]$

S = diagonal (signature) matrix with as many ± 1's as P(0) has positive and negative eigenvalues.

Then we can factor (nonuniquely)

$$\dot{P}(0) = L_0 S L_0'$$

where L_0 is an $n \times \alpha$ matrix, and S is $\alpha \times \alpha$. Show that the gain function K(•)

can be computed via the equations

$$\dot{K}(t) = L(t)SL'(t)H' \ , \quad K(0) = \Pi_0 H'$$
$$\dot{L}(t) = (F - K(t)H)L(t) \ , \quad L(0) = L_0$$

Exercise 9.2. Nonuniqueness of Initial Factorization [20]

a) Suppose $A(\cdot)$ is a nonsingular differentiable matrix such that

$$A(t)SA'(t) = S \ , \quad \text{all } t \ .$$

Then show that there exists $a(\cdot)$ such that

$$\dot{A}(t) = A(t)a(t) \ , \quad a(t)S + Sa'(t) \equiv 0$$

b) Use the above result to show that the Chandrasekhar equations of Exercise 9.1. can be written as

$$\dot{K} = L(t)SL'(t)H' \ , \quad K(0) = \Pi_0 H'$$
$$\dot{L}(t) = (F - K(t)H)L(t) + L(t)a(t) \ , \quad L(0) = L_0 a(0)$$

where $a(\cdot)$ is an arbitrary skew-S symmetric matrix (i.e., $a(t)S + Sa'(t) \equiv 0$).

c) Show how this flexibility in choosing $a(\cdot)$ can be used to make $L(\cdot)$ triangular.

Exercise 9.3. Time-invariant Models [10]

Given a state-space model over (τ, t) with $x(\tau) = 0$, $t > \tau$ let

$$P(t, \tau) = E\, \tilde{x}(t|t)\tilde{x}'(t|t).$$

Show that

a) $$\frac{d}{dt} P(t,\tau) = F(t)P(t,\tau) + P(t,\tau)F'(t) - P(t,\tau)H'(t)H(t)P(t,\tau) + $$
$$+ G(t)Q(t)G'(t), \quad P(\tau,\tau) = 0, \quad t \geqslant \tau$$

b) $$-\frac{d}{d\tau} P(t,\tau) = \Phi(t,\tau)G(\tau)Q(\tau)G'(\tau)\Phi'(t,\tau) ,$$

where $\Phi(.,.)$ is the state-transition matrix of $F(.) - P(.)H'(.)H(.)$.

c) Use these results to obtain Chandrasekhar equations for time-variant models.

REFERENCES

[1] J.L. Casti, R.E. Kalaba and V.K. Murthy, " A new initial-value method for on-line filtering and estimation," **IEEE Trans. Inform. Theory**, vol. IT-18, pp. 515-518, July 1972.

[2] S. Chandrasekhar, "On the radiative equilibrium of a stellar atmosphere, Pt XXI," **Astrophys. J.**, vol. 106, pp. 152-216, 1947; Pt. XXII, ibid, vol. 107, pp. 48-72, 1948.

[3] V.A. Ambartsumian, "Diffuse reflection of light by a foggy medium", **Dokl. Akad. Sci. SSSR**, vol. 38, pp. 229-322, 1943.

[4] V.V. Sobolev, **A Treatise on Radiative Transfer**, New York: Van Nostrand, 1963 (translated from Russian original, 1958).

[5] T. Kailath, "Some Chandrasekhar-type algorithms for quadratic regulator problems", in **Proc. IEEE Conf. Decision and Control and 11th Symp. Adaptive Processes,** Dec. 1972, pp. 219-223. Also Technical Report, Computing and Control Dept., Imperial College, London, Aug. 1972.

[6] T. Kailath, "Some new algorithms for recursive estimation in constant linear systems", **IEEE Trans. Inform. Theory**, vol. IT-19, pp. 750-760, Nov. 1973.

[7] T. Kailath, M. Morf and G.S. Sidhu, "Some new algorithms for recursive estimation in constant discrete-time linear systems", in **Proc. 7th Princeton Symp. Information and System Science**, 1973. See also **IEEE Trans. Automat. Contro.**, vol. AC-19, pp. 315-323, Aug. 1974.

[8] B. Dickinson, M. Morf and T. Kailath, "Canonical matrix fraction and state space description for deterministic and stochastic linear systems", **IEEE Trans. on Automatic Control,** Special Issue on System Identification and Time-Series Analysis, vol. AC-19, No. 6, pp. 656-667, Dec. 1974.

[9] M. Morf and T. Kailath, "Square-root algorithms for linear least-squares estimation and control", **IEEE Trans. Auto. Cont.**, vol. AC-20, pp. 487-497, Aug. 1975.

[10] L.Ljung and T. Kailath, "A scattering theory framework for fast least-squares algorithms", in Multivariate Analysis IV, ed by P.R. Krishnaiah, North-Holland, Amsterdam, 1977. [Presented at Symposium, June 1975].

[11] L. Ljung, T. Kailath and B. Friedlander, "Scattering Theory and Linear Least-Squares Estimation, Part I – Continuous-time Problems, **Proceedings of the IEEE**, vol. 64, pp. 131-139, Jan. 1976.

[12] B. Friedlander, T. Kailath and L. Ljung, "Scattering Theory and Linear Least-squares Estimation", Part II – Discrete-time Problems, J. Framblin Inst., vol. 301, pp. 71-82, Feb. 1976.

[13] G. Verghese, B. Friedlander and T. Kailath, "Scattering Theory and Linear Least-squares Estimation", Part III – The Estimates, IEEE Trans. Automat. Contr., vol. AC-25, pp. 794-802, Aug. 1980.

[14] T. Kailath and L. Ljung, "Efficient Change of Initial Conditions, Dual Chandrasekhar Equations, and Some Applications", **IEEE Trans. Autom. Contr.**, vol. AC-22, pp. 443-447, June 1977.

[15] T. Kailath, A. Vieira and M. Morf, "Orthogonal Transformation, Square-root Implementations of Fast Estimation Algorithms", Int. Symp. on Systems Optim. and Analysis, pp. 81-91, Lecture Notes in Control and Inform. Sciences, vol. 14, New York: Springer-Verlag, 1979.

[16] T. Kailath, B. Lévy, L. Ljung and M. Morf, "The factorization and Representation of Operators in the Algebra Generated by Toeplity Operators", SIAM J. Appl. Math., vol. 37, pp. 467-484, Dec. 1979.

[17] R.S. Bucy and P.D. Joseph, **Filtering for Stochastic Processes with Applications to Guidance.** New York: Wiley, 1968.

[18] R. Bellman, R. Kalaba and J. Wing, "Invariant imbedding and mathematical physics-I: Particle processes", **J. Math. Phys.**, vol. 1, pp. 280-308, 1960.

[19] G.C. Stokes, **Collected Papers,** vol. 2, pp. 91-93, Cambridge Univ. Press, 1883.

[20] M. Morf, B. Lévy and T. Kailath, "Square-root Algorithms for the Continuous-time Least-squares Estimation Problem", IEEE Trans. Automat. Contr., vol. AC-23, pp. 907-911, Oct. 1978.

10. SOME RELATED PROBLEMS

The subject of linear least-squares estimation is a very rich and diverse one, with new results, connections and insights still emerging. A long recent survey [1] of the subject, with over 400 references, still covers only some aspects of the field, and furthermore many additional results have been obtained since its appearance in March 1974, see, e.g., Appendices I and II.

Perhaps the main reason for this vitality is that the subject is intimately connected with he basic structural properties not only of random and deterministic signals, but also of linear and nonlinear systems. Therefore least-squares estimation has ramifications in a very large number of other problems, both deterministic and stochastic.

We very briefly note some of these here.

Dual Quadratic Regulator Problems

In 1960, Kalman noted that the state-space quadratic regulator problem of choosing the control $u(\cdot)$ in

$$\dot{x}(t) = F\,x(t) + G\,u(t), \qquad x(t_0) = x_0$$

so as to minimize the quadratic cost

$$J = x'(t_f)Q_f x(t_f) + \int_{t_0}^{t_f} [x'(t)Q(t)x(t) + u'(t)R(t)u(t)]\,dt$$

could by the associations,

$$F \leftrightarrow F', \quad G \leftrightarrow H', \quad Q \leftrightarrow R$$
$$\Pi_0 \leftrightarrow Q_f, \quad t - t_0 \leftrightarrow t_f - t$$

be mapped into the linear least-squares estimation problem of Section VI. Therefore all results for one problem have analogs in the other so called "dual" problem. These parallels have been fairly thoroughly worked out for "nonsingular" problems in which the weighting matrix $R(t)$ is strictly positive-definite, but more such work needs to be done in the singular case. Some useful references are [2] - [4]. Not unexpectedly, the results in the singular case are closely related to the basic structural properties of linear systems; some of these relations are pursued in [5] - [8], but again there is room for much further work.

As evidence for this, we might note that the above 'dual' associations are not the only ones possible – see, e.g., [9] - [10].

A widely known result in control is the so-called separation theorem which says that when the states of the control system are not directly accessible because of random system inputs and observation noise, then with the estimation model of Section VI, the least-squares estimate of the state can be used in place of the unavailable true states in solving the quadratic problem, where now the expected value of J is minimized rather than J itself. This so-called separation theorem is fairly obvious for state-space models if the innovations model is used to replace the given state-model. More general analyses are provided by Lindquist [11] and Davis [12].

Detection of Gaussian Signals

The separation of estimation and control that we noted above has also been encountered in signal detection problems where the optimum receiver for random signals in white Gaussian noise is identical to the receiver for known signals with the least-squares estimate of the signal being used in place of the unavailable known signal. This result means that many estimation results can be used in detection theory and these possibilities are explored in [13] - [15]. A particular consequence of the study of fast linear estimation algorithms has been a new time-invariant implementation, based on the results described in Appendix I – see [15] - [17].

Miscellaneous

The results and concepts of linear least-squares estimation are useful in many other problems as well. We shall list some of these here, along with some references chosen in part because they are good sources for further exploration and reading:

1. Matrix Factorization and Inversion - [18], [19], [20], [21], [22], [23]
2. Solution of Integral Equations - [24], [25], [26], [27], [28], [29]
3. Stability of Linear Systems - [30], [31], [32]
4. Inverse Scattering Problems - [33], [34], [35], [36], [37]
5. Modeling of Stochastic Systems - [38], [39], [40], [41], [42], [43]
6. Network and System Theory - [44], [45], [46], [47], [48], [49]

7. Spline Approximation by Spline Functions - [50], [51], [52], [53]
8. Spectral Estimation - [54], [55], [56], [57], [58]
9. Applications in System Identification - [59], [60], [61], [62], [63], [64], [65]
10. Applications in Data Communications - [66], [67], [68], [69], [70], [71], [72]
11. Applications in Signal Processing - [73], [74], [75], [76], [77], [78] [79].

REFERENCES

[1] T. Kailath, "A View of Three Decades of Linear Filtering Theory", **IEEE Trans. on Inform. Thy., IT-20,** no. 2, pp. 145-181, March 1974.

[2] D. Jacobson, "Totally Singular Quadratic Minimization", **IEEE Trans. on Autom., Contr., AC-16,** pp. 651-658, 1971.

[3] N. Krasner, "Separable Kernels in Linear Estimation and Control", Ph.D. Dissertation, Dept. of Elec. Engg., Stanford Univ., Stanford, CA, May 1974.

[4] D.J. Clements and B.D.O. Anderson, **Singular Optimal Control: The Linear-Quadratic Problem", Lecture Notes in Control, vol. 5,** Springer-Verlag, New York, 1978.

[5] L. Silverman, "Discrete Riccati Equations: Alternative Algorithms, Asymptotic Properties and System Theory Interpretations", in **Advances in Control, vol. 12,** C.T. Leondes, ed., pp. 313-3386, Academic Press, New York, 1976.

[6] B.P. Molinari, "A Strong Controllability and Observability in Linear Multivariable Control", **IEEE Trans. on Autom. Contr., AC-21,** pp. 761-764, October 1976.

[7] G. Verghese, "Infinite-Frequency Behaviour in Generalized Dynamic Systems", Ph.D. Dissertation, Dept. of Elec. Engg., Stanford Univ., Stanford, CA, December 1978.

[8] A. Emami-Naeini, "Application of the Generalized Eigenstructure Problem to Multivariable Systems and the Robust Servomechanism for a Plant which Contains an Implicit Internal Model", Ph.D. Dissertation, Dept. of Elec., Engg., Stanford Univ., Stanford, CA, May 1981.

[9] W.L. Chan, "Variational Dualities in the Linear Regulator and Estimation Problems", **J. Inst. Math. Appl., 18,** pp. 237-248, October 1976.

[10] U.B. Desai and H.L. Weinert, "Generalized Control-Estimation Duality and Inverse Projections", **Proc. 1979 Conf. on Information Sciences and Systems,** pp. 115-120, John Hopkins Univ., Maryland, March 1979.

[11] A. Lindquist, "Optimal Control of Linear Stochastic Systems with Applications to Time Lag Systems", **Information Sci., 5, pp. 81-126, 1973.**

[12] M.H.A. Davis **Linear Estimation and Stochastic Control,** Halsted Press, New York, 1977.

[13] T. Kailath, "A General Likelihood Ratio Formula for Random Signals in Noise", **IEEE** Trans. on Inform. Thy., IT-15, no. 3, pp. 350-361, May 1969.

[14] T. Kailath, Likelihood Ratios for Gaussian Processes, **IEEE Trans. on Inform. Thy., IT-16,** no. 3, pp. 276-288, May 1970.

[15] M.H.A. Davis and E. Andreadakis, "Exact and Approximate Filtering in Signal Detection: An Example", **IEEE Trans. on Inform. Thy., IT-23,** pp. 768-772, Nov. 1977.

[16] T. Kailath, B. Levy, L. Ljung and M. Morf, "Fast Time-Invariant Implementations of Gaussian Signal Detectors", **IEEE Trans. Inform. Thy., IT-24,** no. 4, pp. 469-477, July 1978.

[17] B. Levy, T. Kailath, L. Ljung and M. Morf, "Fast Time-Invariant Implementations for Linear Least-Squares Smoothing Filters", **IEEE Trans. Autom. Contr., 24,** no. 5, pp. 770-774, Oct. 1979.

[18] T. Kailath, A. Vieira and M. Morf, "Inverses of Toeplitz Operators, Innovations, and Orthogonal Polynomials", **SIAM Rev., 20,** no. 1, pp. 106-119, Jan. 1978.

[19] B. Friedlander, M. Morf, T. Kailath and L. Ljung, "New Inversion Formulas for Matrices Classified in Terms of their Distance from Toeplitz Matrices", **Linear Algebra and Its Appls.,** pp. 31-60, Oct. 1979.

[20] M. Morf, "Doubling Algorithms for Toeplitz and Related Equations", **Proc. Int.'l. Conf. on Acoustics, Speech and Signal Proc.,** pp. 954-959, Denver, Co., 1980.

[21] J.M. Delosme, "Fast Algorithms for Finite Shift-Rank Processes", Ph.D. Dissertation, Stanford Univ., Dept. of Elec. Engg., Stanford, CA, Dec. 1981.

[22] H. Lev-Ari, "Modeling and Parametrization of Nonstationary Stochastic Processes", Ph.D. Dissertation, Stanford Univ., Dept. of Elec., Engg., Stanford, CA, 1982.

[23] B. Porat, "Contributions to the Theory and Applications of Lattice Filters", Ph.D. Dissertation, Stanford Univ., Dept. of Elec. Engg., Stanford, CA, Dec. 1982.

[24] T. Kailath, "Equations of Wiener-Hopf Type in Filtering Theory and Related Applications", in **Norbert Wiener: Collected Works, vol. III, edited by P.R. Masani** MIT Press, 1982.

[25] C.W. Helstrom, "Solution of the Detection Integral Equation for Stationary Filtered White Noise", **IEEE Trans. on Inform. Thy., IT-11, pp. 335-339, 1965.**

[26] T. Kailath, "Fredholm Resolvents, Wiener-Hopf Equations and Riccati Differential Equations", **IEEE Trans. on Inform. Thy., IT-15,** no. 6, pp. 655-672, November 1969.

[27] B. Anderson and T. Kailath, "Some Integral Equations with Nonsymmetric Separable Kernels", **SIAM J. Appl. Math., 20,** no. 4, pp. 659-669, June 1971.

[28] T. Kailath, R. Geesey and H. Weinert, "Some Relations Between RKHS Norms, Innovation Representations and Fredholm Equations", **IEEE Trans. on Inform. Thy., IT-18,** no. 3, pp. 341-348, May 1972.

[29] T. Kailath, L. Ljung and M. Morf, "Generalized Krein-Levinson Equations for Efficient Calculation of Fredholm Resolvents of Nondisplacement Kernels", in **Topics in Functional Analysis,** pp. 169-184, I.C. Gohberg and M. Kac, eds., Academic Press, New York, 1978.

[30] A.J. Berkhout, "Stability and Least-Squares Estimation", **Automatica, 11,** pp. 633-638, 1975.

[31] J.C. Willems, "Dissipative Dynamical Systems I: General Theory; II: Linear Systems with Quadratic Supply Rates", **Archive for Rational Mechanics and Analysis, 45,** pp. 321-343, 1972.

[32] A. Vieira and T. Kailath, "On Another Approach to the Schur-Cohn Criterion", **IEEE Trans. Circuits and Systems, 24,** pp. 218-220, April 1977.

[33] F.J. Dyson, "Old and New Approaches to the Inverse Scattering Problem", in **Studies in Mathematical Physics, Essays in Honor of Valentine Bargmann,** pp. 151-167, E.H. Lieb, B. Simon and A.S. Wightman, eds., Princeton Univ. Press, 1976.

[34] B.D. Anderson and T. Kailath, "Fast Algorithms for the Integral Equations of the Inverse Scattering Problems", **Integral Eqs. and Operator Thy., 1,** no. 1, pp. 132-136, 1978.

[35] K.M. Case, "Inverse Scattering, Orthogonal Polynomials and Linear-Estimation", in Topics in Functional Analysis, pp. 25-43, I. Gohberg and M. Kac, eds., Academic Press, New York, 1978.

H. Dym, "Applications of Factorization Theory to the Inverse Spectral Problem", **Proc. Int'l. Symp. on Math. Thy. of Networks and Systems,** pp. 188-193, Delft, Holland, July 1979. Published by Western Periodicals Co., North Hollywood, CA.

[37] P. Dewilde, J. Fokkema and I. Widya, "Inverse Scattering and Linear Prediction: The Continuous-Time Case", in **Stochastic Systems,** M. Hazewinkel and J.C. Willems, eds., Reidel, 1981.

[38] H. Akaike, "Stochastic Theory of Minimal Realization", **IEEE Trans. Autom. Contr., AC-19,** pp. 667-674, Dec. 1974.

[39] H. Akaike, "Markovian Representation of Stochastic Processes by Canonical Variables", **SIAM J. Contr., 13,** pp. 162-173, 1975.

[40] B.D.O. Anderson and T. Kailath, "The Choice of Signal Process Models in Kalman-Bucy Filtering", **J. Math. Anal. and Appls., 35,** pp. 659-668, Sept. 1971.

[41] A. Lindquist and G. Picci, "On the Stochastic Realization Problem", **SIAM J. Contr. Optim., 17,** pp. 365-389, 1979.

[42] G. Ruckebusch, "On the Theory of Markovian Representation", in **Lecture Notes in Mathematics, 695,** pp. 77-88, Springer-Verlag, New York, 1978.

[43] T. Kailath and L. Ljung, "Strict-Sense State-Space Realizations of Nonstationary Gaussian Processes", to appear.

[44] B.D.O. Anderson and S. Vongpanitlerd, **Network Analysis and Synthesis, A Modern Systems Theory Approach,** Prentice-Hall, New Jersey, 1973.

[45] P. Dewilde, A. Vieira and T. Kailath, "On a Generalized Szego-Levinson Realization Algorithm for Optimal Linear Prediction Based on a Network Synthesis Approach", **IEEE Trans. Circuits and Systems, CAS-25,** pp. 663-675, Sept. 1978.

[46] B.D.O. Anderson and T. Kailath, "Forwards and Backwards Models for Finite-State Markov Processes", **Adv. Appl. Probl., 11,** pp. 118-133, 1979.

[47] P. Dewilde and H. Dym, "Schur Recursions, Error Formulas and Convergence of Rational Estimators for Stationary Stochastic Sequences", **IEEE Trans. Inform. Thy.,** IT-pp. 446-461, 1981.

[48] Y. Genin and P. Van Dooren, "On Σ -Lossless Transfer Functions and Related Questions", Philips Res. Lab. Tech. Rept. R447, Brussels, Belgium, Nov. 1980.

[49] D.Z. Arov, "Passive Linear Stationary Dynamic Systems", **Siberian Mat. J., 20.,** pp. 149-162, 1979.

[50] G. Kimeldorf and G. Wahba, "Spline Functions and Stochastic Processes", **Sankhya, 132,** pp. 173-180, 1970.

[51] H.L. Weinert and T. Kailath, "Stochastic Interpretations and Recursive Algorithms for Spline Functions", **Annals of Stat., 2,** pp. 787-794, 1974.

[52] H.L. Weinert and G.S. Sidhu, "A Stochastic Framework for Recursive Computation of Spline Functions, Pt. I, Interpolating Splines", **IEEE Trans. Inform. Thy.,, IT-24,** pp. 45-50, Jan. 1978.

[53] R.J.P. de Figueireido and A. Caprihan, "An Algorithm for the Generalized Smoothing Spline with Application to System Identification", **Proc. 1977 Conf. Inform. Sciences and Systems,** John Hopkins Univ., Baltimore, MD, April 1977.

[54] E. Parzen, "Multiple Time-Series Modeling", in **Multivariate Analysis – II,** P.R. Krishnaiah, ed., pp. 389-409, Academic Press, New York, 1969.

55 J.P. Burg, "Maximum Entropy Spectral Analysis", Ph.D. Dissertation, Dept. of Geophysics, Stanford Univ., Stanford, CA, 1975.

[56] M. Morf, A. Vieira, D. Lee and T. Kailath, "Recursive Multichannel Maximum Entropy Spectral Estimation", **IEEE Trans. Geoscience Elect., GE-16,** no. 2, pp. 85-94, April 1978.

[57] D.G. Childers, ed., **Modern Spectrum Analysis,** IEEE Press, New York, 1978.

[58] S. Haykin, ed., **Nonlinear Spectral Analysis,** Sprjnger-Verlag, New York, 1979.

[59] K.J. Astrom and P. Eykhoff, "System Identification - A Survey", **Automatica, 7,** pp. 123-162, 1971.

[60] M. Morf, T. Kailath and L. Ljung, "Fast Algorithms for Recursive Identifcation", **Proc. 1975 IEEE Conf. on Decision and Contr.,** pp. 916-921, Florida, Dec. 1976.

[61] V. Solo, "Time Series Recursions and Stochastic Approximation", Ph.D. Dissertation, Dept. of Statistics, Australian National Univ., Camberra, Australia, 1978.

[62] R. Isermann, ed., Special Issue on Identification and System Parameter Estimation, **Automatica, 17,** pp. 1-259, Jan. 1981.

[63] V. Strejc, "Trends in Identification", **Automatica, 17,** pp. 1-22, Jan 1981.

[64] P. Young, "Parameter Estimation for Continuous-Time Models – A Survey", **Automatica, 17,** pp. 23-40, Jan. 1981.

[65] L. Ljung and T. Soderstrom, **Theory and Practice of Recursive Identification,** MIT Press, Cambridge, MA, 1982.

[66] D. Messerschmitt, "A Geometric Theory of Intersymbol Interference, Pt. I: Zero-Forcing and Decision-Feedback Equalization", **Bell Syst. Tech. J.,** pp. 1483-1519, 1973.

[67] D. Godard, "Channel Equalization Using a Kalman Filer for Fast Data Transmission", **IBM J. Res. Develop., pp. 267-273, May 1974.**

[68] D.D. Falconer and L. Ljung, "Application of Fast Kalman Estimation to Adaptive Equalization", **IEEE Trans. Communications, COM-26,** pp. 1439-1446, Oct. 1978.

[69] L.E. Franks, ed., **Data Communication,** Benchmark Papers in Electrical Engineering and Computer Science, Halsted Press, New York, 1974.

[70] E.H. Satorius and J.D. Pack, "Application of Least-Squares Lattice Algorithms to Adaptive Equalization", **IEEE Trans. on Communication, COM-29,** pp. 136-142, Feb. 1981.

[71] M.S. Mueller and J. Salz, "A Unified Theory of Data-Aided Equalization" **Bell Syst. Tech. J.,** 1981.

[72] V.U. Reddy, A.M. Peterson and T. Kailath, "Application of Modified Least-Squares Algorithms to Adaptive Echo Cancellation", Proc. Int'l. Symp. on Microwaves and Communication, Kharagpur, INDIA, Dec. 1981.

[73] L.C. Wood and S. Treitel, "Seismic Signal Processing", **Proc. IEEE, 63,** pp. 649-661, A April 1975.

[74] J.D. Markel and A.H. Gray, **Linear Prediction of Speech,** Springer-Verlag, New York, 1976.

[75] J. Makhoul, "Linear Prediction: A Tutorial Review", **Proc IEEE, 63,** pp. 561-580, April 1975.

[76] M. Morf and D.T. Lee, "Recursive Least-Squares Ladder Forms for Fast Parameter Tracking", **Proc. 1978 IEEE Conf. on Decision and Contr.,** pp. 1362-1367, San Diego, CA, Jan. 1979.

[77] S. Wood and M. Morf, "A Fast Implementation of a Minimum Variance Estimator for Computerized Tomography Image Reconstruction", **Trans. Bio-Medical Engineering, BME-28,** pp. 56-88, Feb. 1981.

[78] S.R. Parker and L.J. Griffiths, eds., Joint Special Issue on Adaptive Signal Processing, **IEEE Trans. on Acoustics, Speech and Signal Processing, ASSP-29,** June 1981.

[79] A. Benveniste and P. Dewilde, eds., Fast Algorithms for Linear Dynamical Systems, published by INRIA/CNRS, Paris, France, 1982. 48

APPENDIX I

SOME NEW RESULTS AND INSIGHTS IN LINEAR LEAST–SQUARES ESTIMATION THEORY

Thomas Kailath(*)
Department of Electrical Engineering
Stanford University
Stanford, California 94305

Reprinted, with corrections, from Proceedings of the 1975 IEEE–USSR Joint Workshop on Information Theory, pp. 97-104, Moscow, Dec. 1975.

(*) This work was supported in part by the Air Force Office of Scientific Research, A.F. Systems Command under Contract AF44-C-0068, in part by the Joint Services Electronics Program under Contract N00014-67-0112-0044, and in part by the National Science Foundation under Contract NSF-Eng 75-18952.

Abstract

We give an outline of a new trend in least-squares estimation theory, and more generally in linear systems theory, away from state-space models and back to input-output descriptions. This has grown out of the realization that the computationally attractive recursive solutions associated with state-space models can also be achieved for input-output models characterized in terms of their "distance from stationarity". When additional state-space structure is imposed on such models, the input-output recursions can be reduced to the known state-space algorithms. The underlying ideas here are also useful in a variety of related problems, both deterministic and stochastic.

I. Introduction

The increasing use of dynamical (state-space) models has been one of the most significant developments of the last decade in the general field of information processing, replacing the previously more-common input-output models [1]. However armed with the insights gained from the state-space theory, there has recently been an increasing return to input-output models, which are more readily at hand in many applications. This has been especially true in linear system theory, chiefly under the leadership of Rosenbrock [2] (see also Popov [3]), and in the theory of linear least-squares estimation. The major initial contributions to the latter subject were made by mathematicians (especially Gauss, Kolmogorov, Krein and Wiener), but much of the recent development has been made in the engineering literature. Perhaps not surprisingly, one of the major contributions that engineers have made to this theory is the realization of the close interplay between the theory of random processes and the theory of deterministic linear systems. Some very early and by now widely appreciated examples of such a fruitful interplay are Nyquist's theorem on thermal noise, and the analysis of shot noise (Campbell's theorem, filtered Poisson processes, etc.). In least-squares estimation theory, the value of the interplay was strikingly brought out by the linear-filter interpretations of Wiener's results by Bode and Shannon [4] and Zadeh and Ragazzini [5].

In Section II of this paper, we shall try to explain how the natural progress of linear system theory led to the emergence of state-space models in linear estimation theory. The theory seemed to have reached definitive form with a voluminous literature on the Riccati-equation of the state-space based Kalman filter

[6-8], but in 1972 a short note by Casti, Kalaba and Murthy [9] drew the author's attention to some early results (ca. 1945) of Ambartzumian [10] and Chandrasekhar [11] on finite-interval Wiener-Hopf equations for stationary processes. This led to the discovery [12] of Chandrasekhar-type equations for constant-parameter state-space models. A significant result of [12] was that these equations were useful for a class of nonstationary processes,with the degree of usefulness being determined by a somewhat mysteriously defined parameter α (cf. Section III), whose significance has been a matter of some debate [13]. Next a study with M. Morf of the discrete versions of the Chandrasekhar-type equations of [12] showed them to be closely related to his results (ca. 1970)[14] on fast algorithms for the triangular-(Cholesky-) factorization of Toeplitz covariance matrices, to the results of Levinson[15] on fast algorithms for solving linear equations with Toeplitz coefficients matrices, and to the closely related results of Trench [16] and Gohberg and Semencul [17-18] on some explicit formulas for Toeplitz inverses. The author then recognized that the continuous-time analog of the Trench formula was the famous Sobolev identity[19] of radiative-transfer theory and that both these formulas were just another expression of the very old Christoffel-Darboux formulas for orthogonal polynomials [20-22]. It had been known for some time (see, e.g., [23]-[24]) that the Levinson recursions were just those of the Szegö orthogonal polynomials on the unit circle [20-22]; moreover, the continuous analogs of these recursions were also known, having been dicovered by Krein in 1955 [25] (see also Akhiezer [26]). A combination of these results with some extensions of the Chandrasekhar-type equations obtained by the use of a scattering-theory framework (cf. [27]-[29]) led to the realization [30-31] that generalized Krein-Levinson recursions could be obtained for the estimation of arbitrary nonstationary processes, classified in terms of their "distance from stationarity". These results are described in Section IV, while in Section V we show how by imposing additional (constant-parameter) state-space-like structure the Krein-Levinson equations can be reduced to the Chandrasekhar-type equations of Section III. Furthermore the mysterious parameter α appearing in the Chandrasekhar-type equations turns out to be just the distance from stationarity of the signal process. It also turns out that this concept coincides in the discrete case with a notion of "shift-low-rank" of certain classes of matrices (sums of products of Toeplitz and Hankel matrices) studied by Morf (see [14]) ; the discrete-time elaboration of these ideas is given in [32]-[33].

Thus we have come almost a full circle in the theory, tracing a path from stationary processes to Riccati-type equations, then Chandrasekhar-type equations, then Krein-Levinson-type equations and back to the Chandrasekhar-type equations. Actually the general Riccati-type equations for completely time-variant models can also be imbedded in some input-output recursions that exploit a different kind of property than "distance from stationarity", and we make a brief reference to this as well. The message, however, is that the last word has still not been said on this topic and that there is still room for the introduction and exploitation of other kinds of structures.

Some readers might feel that undue attention is being paid to a very narrow topic. This may be so, but on the other hand some comfort may also be drawn from the fact that a partial list of other problems to which the theory of linear least-squares estimation is closely related includes the detection of Gaussian signals in Gaussian noise, quadratic control theory, computation of spline functions, Hurwitz factorization of scalar and matrix polynomials, solution of Fredholm integral equations and linear two-point boundary value problems, minimal design problems in multivariable system identification. References to recent work in some of these areas are briefly noted in the concluding Section VI.

II. The Path to State-Space Models and Riccati-Type Equations

In the introduction, we referred to the fruitful interplay between the theories of linear systems and of random processes. The first connections were between the theory of a **stationary** (second-order) processes and stable linear **time-invariant** filters excited at $t = -\infty$, so that all (nonstationary) transients could be neglected. Furthermore, for reasons of mathematical convenience and practical utility, attention was largely restricted to processes with rational power spectral densities and linear systems with rational transfer functions. A good account of the theory and applications of the mutually fruitful interaction between such processes and filters can be found in the textbook of Lee [34],who was in fact one of the first, under Wiener's active encouragement, to recognize and exploit these connections. Fourier methods, especially as developed by Wiener, were the natural tool for these studies.

But just as this theory was being put into definitive form, developments in technology (the computer revolution and the advent of satellites and the space age) made some of the assumptions of the older theory obsolete. For example, it was not reasonable to assume that observations began in the remote past, and transient nonstationarities could not be avoided. Therefore, in particular, Fourier methods were not directly applicable and it became almost essential to use time-domain language. Thus the transfer function description

$$\mathcal{H}(s) = \frac{b_1 s^m + \ldots + b_m}{s^n + a_1 s^{n-1} + \ldots + a_n} \tag{1}$$

had to be replaced by the input-output differential equation,

$$y^{(n)}(t) + a_1 y^{(n-1)}(t) + \ldots a_n y(t) = b_1 u^{(m)}(t) + \ldots + b_m u(t), \tag{2}$$

where the finite starting time

$$t \geqslant t_0 > -\infty \tag{3}$$

could be explicitly introduced.

Now, as stated earlier, transients will arises due to the initial conditions

$$y(t_0), y^{(1)}(t_0), \ldots, y^{(n-1)}(t_0) \tag{4}$$

However, actual calculation of the transients, though simple in concept, is algebraically complicated if the form (2) is used. This point can be clearly seen in the explicit but cumbersome expressions to be found in the early papers on this question ; some textbooks attempting to codify these results were those of Solodovnikov [35a], Pugachev [35b] and Laning and Battin [36]. And it is very interesting to see how the latter authors were driven almost inexorably to introduce state-space descriptions of linear systems, where the use of first-order equations and matrix notation combine to give a compact way of handling initial conditions(*). Moreover, simple satellite and rocket tracking problems were also bringing state-equations to the fore (see, e.g., the book of Peterson [37]) ; and via control theory, Bellman and Kalman had begun to emphasize the attractiveness of state-space models for obtaining (computationally efficient) recursive algorithms.

The state-space model has the form

$$\dot{x}(t) = F(t)x(t) + G(t)u(t), \quad x(t_0) = x_0 \tag{5 a}$$

$$z(t) = H(t)x(t) \quad , \tag{5 b}$$

$$y(t) = z(t) + J(t)v(t) \quad , \; t \geqslant t_0 \tag{5 c}$$

(*) As stated in [36, p. 340]: "In effect, the essential difficulties associated with the point $t = 0$ are absorbed in the calculations needed to reduce the N-th order equation to N first-order equations".

where we have the immediate bonuses that without very much extra conceptual burden, we can

i) allow the coefficients to be time-dependent

ii) treat multivariable systems with multiple inputs and outputs.

In particular, to fix the notation, we shall assume that there are n state variables x(.), m inputs u(.), p outputs y(.) so that

$$F(.) \text{ is } n \times n, \quad G(.) \text{ is } n \times m$$

$$H(.) \text{ is } p \times n, \quad J(.) \text{ is } p \times p \ .$$

We now add the assumptions that u(.) and v(.) are zero-mean white-noise processes with

(6a) $$E\, v(t)v'(s) = I\, \delta(t-s), \quad E\, u(t)u'(s) = Q(t)\delta(t-s)$$

(6b) $$E\, u(t)v'(s) = C(t)\delta(t-s)$$

The initial condition is assumed to be random with

(6c) $$E\, x_0 = 0, \quad E\, x_0 x_0' = \Pi_0, \quad E\, u(t)x_0' \equiv 0.$$

Finally, knowledge of the matrices F(.), G(.), H(.), J(.), Q(.), $\Pi_0(.)$ is also assumed and for simplicity we shall further assume that

(6d) $$C(.) \equiv 0 \quad \text{and} \quad J(.) \equiv I\ .$$

Then

(7) $$\hat{x}(t) = \text{the linear least-squares estimate of } x(t) \text{ given past observations } \{y(s),\ s < t\}$$

can be calculated by the celebrated Kalman-Bucy(*) equations [7]

(8a) $$\dot{\hat{x}}(t) = F(t)\hat{x}(t) + K(t)[y(t) - H(t)\hat{x}(t)]\,, \quad \hat{x}(0) = 0$$

(8b) $$\hat{z}(t) = H(t)\hat{x}(t)\,,$$

(9) $$K(t) = P(t)H'(t)\,,$$

(*) It may be noted that the formulas (8)-(11) were actually first obtained by Stratonovich [8] as a special (linearized) form of some general nonlinear estimation equations; however, the above authors were more alert in recognizing and emphasizing the wide scope of these results.

$$P(t) = E\,\tilde{x}(t)\tilde{x}'(t)\,, \quad \tilde{x}(t) = x(t) - \hat{x}(t) \tag{10}$$

where the n x n error-variance matrix P(.) obeys a nonlinear matrix Riccati differential equation

$$\dot{P}(t) = F(t)P(t) + P(t)F'(t) + G(t)Q(t)G'(t) - K(t)K'(t), \quad t \geq t_0 \tag{11a}$$

with initial condition

$$P(t_0) = \Pi_0 = E\,x_0 x_0' \,. \tag{11b}$$

This initial value differential equation can be solved iteratively on an analog or digital computer, and so can the linear differential equation for $\hat{x}(.)$. No data-storage is required, the estimates being recursively updated as each new data element comes in.

The Kalman-Bucy filter has now been so widely used and such a mystique has grown up around it that it seems pointless to look for other techniques. But (with some hindsight) one can think of at least three reasons for doing so :

i) When n is large (say greater than 50 or 100) we have to solve n^2 (or $n^2/2$) simultaneous nonlinear equations to define P(.) and this can be computationally burdensome.

ii) While in one sense it is a great strength of the Riccati-based Kalman-Bucy filter that it holds equally for time-variant and time-invariant models, in another sense this is a weakness because one should be able to exploit the time-invariance to obtain some computational simplifications.

iii) If we only have an external description, this has to be converted into a state-space description and this is not always easy.

We shall now describe a method that begins to address the problems i) and ii) using Chandrasekhar-type equations[12] in place of the Riccati equation. Then we shall indicate how this can lead us to a way of handing problem iii) that will provide a way of exploiting any structure that may be present in the external description. If in particular we know that there is an underlying state-space model, then we shall see that our way of using the external description will reduce naturally to the Chandrasekhar-type solution.

III. Chandrasekhar-Type Equations and the Path Back to Input-Output Models

If the state-space model parameters F(.), G(.), H(.), Q(.) are constant then it was shown in [12] that the "gain" function K(.) to be used in the estimator equation (8) could be found via the "Chandrasekhar-type" equations.

(12a) $$\dot{K}(t) = L(t)\Lambda L'(t)H' \quad , \quad K(t_0) = \Pi_0 H'$$

(12b) $$\dot{L}(t) = [F - K(t)H]L(t) \quad , \quad L(t_0) = L_0$$

where L(.) is an n x α matrix and

(13a) $$\alpha = \text{rank } \dot{P}(0) = \text{rank } [F\Pi_0 + \Pi_0 F' + GQG' - \Pi_0 H' H\Pi_0]$$

(13b) Λ = an α x α "signature" or "inertia" matrix of $\dot{P}(0)$

(13c) L_0 is any matrix such that $L_0 \Lambda L_0' = \dot{P}(0)$.

P(.) is the error-variance matrix, which obeys the Riccati equation (11) in the Kalman-Bucy filter ; but note that we do not need to compute P(.) in the present algorithm, though it may be determined via $\dot{P}(t) = L(t)\Lambda L'(t)$.

Since we only have $n(p + \alpha)$ variables in the equations (12) as compared to $n^2/2$ in the Riccati equation (11) there can be considerable computational saving whenever $p \ll n$, $\alpha \ll n$. In fact, significant benefits have already been demonstrated in an air-pollution study [38] where n = 500 and the discrete-time analog [39] of the equations (12) was used to obtain a reduction in computational complexity over the Riccati equation of 0(125,000,000) computations/step to 0(125,000) computations/step. However from (13a) one can readily conclude that α would be equal to n for almost all initial conditions. Therefore it is important to try to find special cases in which α is low and more generally to explore the meaning of such cases.

To aid in this analysis, we note first that the covariance function of the signal process z(.) generated by a time-invariant state-space model (5) can be written

(14) $$E\, z(t)z'(s) = K(t, s) = H e^{F(t-s)} N(s), \quad t \geqslant s$$

where

(15) $$N(t) = \Pi(t)H', \quad \Pi(t) = E\, x(t)x'(t)$$

and

(16) $$\dot{\Pi}(t) = F\,\Pi(t) + \Pi(t)F' + GQG', \quad \Pi(t_0) = \Pi_0.$$

Now we can see that if F is stable, i.e., its eigenvalues obey

$$\mathrm{Re}\{\lambda_i(F)\} < 0 \ , \tag{17}$$

then

$$\dot{\Pi}(t) \to 0 \ \text{and} \ \Pi(t) \ \to \ \overline{\Pi} \ \text{as} \ t \to \infty \ ,$$

where $\overline{\Pi}$ is the unique nonnegative-definite solution of the Lyapunov equation

$$F \ \overline{\Pi} \ + \ \overline{\Pi} \, F' + \ GQG \ = \ 0 \ . \tag{18}$$

In this case

$$N(t) \ = \ \overline{\Pi} \, H', \tag{19}$$

and the process z(.) is (wide-sense) stationary.

We see moreover that this effect can also be ensured by the special choice of initial condition

$$\Pi_0 \ = \ \overline{\Pi} \ , \tag{21a}$$

and that in this case (cf. (13))

$$\alpha = \ \mathrm{rank} \ \{ -\overline{\Pi} \, H' H \, \overline{\Pi} \} \leqslant \ \min \, (n, \, p) \ . \tag{22}$$

Thus in the important classical case of estimation of scalar (p = 1) stationary processes observed over a finite interval, we have $\alpha = 1$ and thus a substantial potential saving over use of the Riccati equation. Furthermore in view of the great attention devoted to kernels of the type K(t – s) in the mathematical and physical literature, it may not surprise the reader that versions of the equations (12) were first obtained in 1947 by S. Chandrasekhar in the theory of radiative transfer [11a, Ch. 8] . [Chandrasekhar's work relied heavily on that of V.A. Ambartzumian [10] who first introduced the functions K(.) and L(.), but in a different integral form.]

Stationary processes can be easily specified without recourse to state-space models, and the studies of Ambartzumian and Chandrasekhar were for scalar kernels of the form

$$K(t, \, s) \ = \ \int_0^1 e^{-\alpha |t-s|} w(\alpha) \, d\alpha \ .$$

The derivations of these authors used heavily the natural shift-invariance properties of such "displacement" or "convolution" or "Toeplitz" kernels. Their ideas were picked up by Bellman and Kalaba in the early sixties and extensively developed by them and others under the rubric "invariant imbedding" (see, e.g., [40]). However, almost all this work has again been exclusively for displacement kernels, because it was not clear how the "invariance" concept could be incorporated into a general nonstationary problem.

A Class of Nonstationary Processes

But here our discovery [12] of the state-space version (12)-(13) of the Chandrasekhar equations showed a way of handling a special, but not uncommon, class of nonstationary processes. The point is, returning to the formulas (14)-(16), that for any choice of initial condition other than $\overline{\Pi}$, or whenever the state-matrix F is unstable so that $\overline{\Pi}$ is not defined, the output of a constant-parameter state-space model will be a nonstationary process. But the striking fact is that even for such nonstationary processes, the Chandrasekhar equations (12)-(13) continue to hold and we can have $\alpha \ll n$. Perhaps the simplest example is provided by the assumption

$$(23) \qquad \Pi_0 = 0\,, \quad \text{so that} \quad \alpha = \text{rank}\ \{GQG'\} \leqslant \min(m, n)$$

so that there can be a substantial advantage if m, the number of inputs to the system, is low.

The fact that time-invariance can be exploited to potential computational advantage has also been demonstrated for other classes of estimation algorithms, especially the so-called square-root array methods [41] .

We shall not describe this here, however, because our goal is to explore further the significance of the special parameter α that we have encountered above.

We first note again that it is reasonable to expect that for almost all Π_0 , the rank α of $\dot{P}(0)$ in (13a) will be n, and therefore of "maximum" complexity. However, this ignores two things. Often the cases of physical interest are those that have some special characteristic and this may lead to low α as in the two examples presented above. The other point is that the number n defines maximum complexity only because we are restricting ourselves to finite-dimensional models. We may note in passing that (analogously to the statements we made on Π_0 and α) one could also say that almost all physical systems are not finite-dimensional. If we consider arbitrary processes, then a finite α is very small for processes with a possibly infinite-dimensional state-space. The finiteness of α in such cases often arises from the finite number of inputs and outputs of the infinite-dimensional state-space model, and this fact indicates that a renewed consideration of input-output system descriptions may be valuable. Of course, there has always also been the point that in many problems, no state-space models are conveniently at hand and one has to work with external descriptions.

The reason for a reconsideration at this point is that we recently recognized that the "fast" Chandrasekhar-type algorithms presented in this section were closely related to some very general results on integral (and matrix) equations with arbitrary (nonstate-space-related) "stationary" or "Toeplitz" or "displacement" kernels.

For example, we were aware that there existed recursive algorithms for solving linear equations with an n x n Toeplitz coefficient matrix in $0(n^2)$ operations, as opposed to the $0(n^3)$ needed for an arbitrary n x n coefficient matrix. In particular we have the Levinson algorithm [15] widely used in geophysical analysis, maximum entropy spectral analysis, speech processing, etc. (see, e.g., [42]-[43]), but not much known in the state-space dominated control iterature. Closely related to the Levinson algorithm(*) is one due to Trench [16] for inverting a Toeplitz matrix. The continuous-time analog of Trench's algorithm was discovered in astrophysics by V.L. Sobolev [19] and then in more general form by Gohberg and Semencul [17]. Soon after the appearance of Levinson's algorithm [15] for solving Toeplitz equations, it was noted by Burg, Robinson, Whittle [23-34] and others that the key recursions in [15], [44] were identical to those of the Szegö orthogonal polynomials on the unit circle. The continuous-time analogs were discovered by Krein [25]; it was then noticed [22] that the Trench and Sobolev formulas were just the Christoffel-Darboux identities for orthogonal polynomials.

However, these results are all for Toeplitz kernels. But we saw in connection with the Chandrasekhar equations that it was possible to extend the potential computational advantages of Toeplitz kernels (stationary processes), to non Toeplitz kernels (nonstationary processes) whose "distance" in some sense from Toeplitz is measured by a parameter α. However, α as defined by (13a) seems to be heavily tied to the special state-space structure and the question arises whether a similar parameter can be defined in more general circumstances.

(*) The Levinson algorithm was rediscovered by Durbin [44] in the problem of fitting autoregressive models to covariance data.

In fact, the answer is yes, as is described in the next two sections.

IV. Recursions for the Impulse Response of the Optimal Filter

Following the lead provided by the Levinson algorithm [15] for stationary processes, we have discovered [30-33] methods for recursively updating the impulse response for estimation of nonstationary processes. In the stationary case, our equations reduce to the Levinson-Szegö recursions in discrete-time, and to the Krein equations in continuous-time. Our nonstationary generalizations depend upon a measure of a "distance from stationarity" of the nonstationary process. This notion can be introduced as follows.

First let us write

$$\lrcorner[K(t,s)] = \left(\frac{\partial}{\partial u} + \frac{\partial}{\partial v}\right) K(u,v)\Big|_{\substack{u=t \\ v=s}} . \tag{24}$$

Then the (first-order) displacement rank of a stochastic process with covariance function K(t,s) will be defined (30) as the smallest number $\tilde{\alpha}$ such that we can write

$$\lrcorner[K(t,s)] = K(t,0)K(0,s) + D'(t)\Lambda D(s) \tag{25}$$

for some p x α matrix $D'(.)$ and some $\alpha \times \alpha$ block-diagonal matrix Λ.

For a stationary process, K(t,s) depends only upon $|t-s|$ and we see that

$$\lrcorner[K(t,s)] = 0 \tag{26}$$

and we can take

$$D'(t) = K(t,0)\,, \quad \Lambda = -I, \quad \alpha = p\,. \tag{27}$$

For nonstationary processes, α will generally be larger than p. The simplest nonstationary process is perhaps the Wiener process, and this has

$$K(t,s) = \min(t,s)\,, \quad \lrcorner[K(t,s)] = 1, \quad \alpha = 1.$$

The so-called Brownian bridge process has

$$K(t,s) = \min(t,s) - ts,$$

$$\lrcorner[K(t, s)] = 1 - t - s\,, \quad \alpha = 2\,.$$

We shall see later (Section V) that a process with an n-dimensional state-space model will have $\alpha \leqslant n$.

Now suppose that we are given an observed process y(.), with covariance function

$$E\, y(t)y'(s) = I\,\delta(t - s) + K(t, s) \tag{28}$$

where K(. , .) has displacement rank α ; we wish to calculate h_{xy} (t,s) so that

$$\hat{x}(t) = \int_0^t h_{xy}(t, s)y(s)\, ds, \quad 0 \leqslant t \leqslant T$$

$$= \text{the linear least-squares estimate of} \tag{29}$$

$$x(t) \text{ given } \{y(s),\ s \leqslant t\}\,.$$

It is easy to see that h_{xy} (t,s) must obey the Wiener-Hopf equation

$$h_{xy}(t, s) + \int_0^t h_{xy}(t, r)K(r, s)\, dr = K_{xy}(t, s), \quad 0 \leqslant s \leqslant t \leqslant T \tag{30}$$

where

$$K_{xy}(t, s) = E\, x(t)y'(s)\,. \tag{31}$$

Some assumptions about the relation between x(.) and y(.) will be necessary to say more about the problem, and we shall assume that

$$\lrcorner[K_{xy}(t, s)] = K_{xy}(t, 0)K(0, s) + D'_{xy}(t)\Lambda D(s) \tag{32}$$

where K(0,s) and D(s) are the same functions as arise in the expression (25) for $\lrcorner[$ K(t,s) $]$. One justification for this assumption (32) is that it is consistent with an assumption often made in least-squares estimation problems that

$$y(t) = Hx(t) + v(t)\,, \tag{33}$$

where v(.) is unit-intensity white noise uncorrelated with x(.). For then

$$K(t, s) = HE[x(t)x'(s)]H' = HK_{xy}(t, s)\,, \tag{34a}$$

and we can take

$$D'(t) = HD'_{xy}(t)\,. \tag{34b}$$

In any case, with the assumptions (25) to (34) it can be shown [31] that

h_{xy} (t,s) can be computed via the so-called extended Krein-Levinson equations

(35) $$h_{xy}(t, s) = C'_{xy}(t, t)\Lambda C_{xy}(t, s)H' , \quad s \leqslant t$$

(36) $$\frac{\partial}{\partial t} C'_{xy}(t, s) = - A'_{xy}(t, s)HC'_{xy}(t, t) , \quad s \leqslant t$$

(37) $$A'_{xy}(t, s) = C_{xy}(t, s)\Lambda C_{xy}(t, t)H'$$

with boundary values determined via the equations

(38) $$h_{xy}(t, 0) = K_{xy}(t, 0) - \int_o^t h_{xy}(t, r) K(r, 0) dr$$

(39) $$C'_{xy}(t, t) = D'_{xy}(t) - \int_o^t h_{xy}(t, r)HD'_{xy}(r) dr$$

(40) $$A'_{xy}(t, 0) = K_{xy}(0, t) - \int_o^t K_{xy}(0, r)HA'_{xy}(t, r) dr$$

In the special case,

$$p = 1 , \quad H \equiv I , \quad K(t, s) = K(t - s)$$

it can be verified that these recursions reduce to the equations first presented by Krein for the continuous analog of the Szegö orthogonal polynomials on the unit circle [25]. The general equations can be solved in various ways, e.g., by discretizing the interval of interest into say N subintervals, and replacing the above differential equations by suitable recursions. It is not hard to see that to determine h_{xy} will require proportional to $p^2 \alpha N^2$ operations ; this can be a substantial savings over the (proportional to) p^3N^3 operations required to solve the usual discretized version of (30), since N usually has to be large to get a good approximation. The degree of saving depends upon the value α- -the lower it is, the closer the process is to a stationary process and the greater the saving. This number α puts enough structure into the nonstationary problem to allow our recursive solution ; as mentioned earlier, until now such solutions have been widely regarded as being conditional on the availability of state-space models.

Actually our recursive solutions have the property that by adding "state-space-like" assumptions, they can be reduced to a generalized form of the

Chandrasekhar-type equations of Section III. This question will be pursued in the next section, at the end of which we shall also remark how the Riccati equations of Section II can be imbedded into an input-output framework [45].

V. Specializing the Input-Output Recursions to the State-Space Case

Let us first assume that

$$y(t) = H(t)x(t) + v(t) \tag{41}$$

$$E\, v(t)v'(s) = I\,\delta(t-s)\,,\quad E\, v(t)x'(s) \equiv 0\,. \tag{42}$$

Then it is easy to check that

$$K(t,s) = H(t)[E\, x(t)x'(s)]H'(s) = H(t)K_{xy}(t,s)\,. \tag{43}$$

Let us now introduce a further assumption about the processes x(.) and y(.), namely that there exists a function F(.) such that

$$\frac{\partial}{\partial t}K_{xy}(t,s) = F(t)K_{xy}(t,s)\;;\; 0 \leqslant s \leqslant t \leqslant T\,. \tag{44}$$

For example, this will be true if K_{xy} (. , .) has a form common in radiative-transfer theory [10-11]

$$K_{xy}(t,s) = \int_0^1 e^{-\alpha|t-s|}\, w(\alpha)\, d\alpha\,. \tag{45}$$

It will also be true when we have a state-space model of the form (5), for then a standard calculation shows that

$$K_{xy}(t,s) = \Phi(t,s)\Pi(s)H'(s)\,,\quad t \geqslant s \tag{46}$$

where

$$\frac{\partial}{\partial t}\Phi(t,s) = F(t)\Phi(t,s) \tag{47}$$

and Π (.) is defined by (16).

Then we can show [31] that

$$\frac{\partial}{\partial t} h_{xy}(t,s) = (F(t) - h_{xy}(t,t)H(t))h_{xy}(t,s)\,, \quad s < t\,. \tag{48}$$

and

$$\begin{aligned}\dot{\hat{x}}(t) &= \frac{\partial}{\partial t}\int_0^t h_{xy}(t,s)y(s)\,ds \\ &= \int_0^t (F(t) - h_{xy}(t,t)H(t)h_{xy}(t,s)y(s)ds + h_{xy}(t,t)y(t) \\ &= F(t)\hat{x}(t|t) + h_{xy}(t,t)(y(t) - H(t)\hat{x}(t|t)\,,\end{aligned} \tag{49}$$

the usual recursive state-estimator equation of Kalman and Bucy [7]. From (49) we note the important fact that the function

$$h_{xy}(t,t) = H_{xy}(t,t;t) = A_{xy}(t\,;t) \tag{50}$$

suffices to specify $\hat{x}(t)$ under the assumptions (41) and (45).

Now if we add the assumption that F(.) and H(.) are **time-invariant**, it turns out [31] that this function can be determined from the equations

$$\frac{d}{dt} h_{xy}(t,t) = C'_{xy}(t,t)\Lambda C_{xy}(t,t)H' \tag{51a}$$

$$\frac{d}{dt} C'_{xy}(t,t) = (F - h_{xy}(t,t)H)C'_{xy}(t,t) \tag{51b}$$

with initial conditions

$$h_{xy}(0,0) = K_{xy}(0,0)\,,\quad C'_{xy}(0,0) = D'_{xy}(0)\,, \tag{52}$$

But these are just the Chandrasekhar-type equations (12)-(13) that we presented earlier in Section III. Here we actually have a somewhat more general result since we have started with the more general assumptions (41) and (45). For example, if a state-model (5) is explicitly assumed, (41) and (45) only require that F(.) and H(.) be time-invariant and not also G(.) and Q(.). Such generalized Chandrasekhar equations were first derived in [29] : the fact that time-invariance of G(.) and Q(.) was not necessary to get fast algorithms was first pointed out in [41].

The significance of the constancy of all parameters in the state model is that we can then obtain a nice formula for the displacement rank.

Thus when F and H are constant, (47) yields

$$\lrcorner K(t,s) = H\ e^{F(t-s)}\dot{\Pi}(s)H' \ , \quad s \leqslant t \tag{53}$$

where

$$\dot{\Pi}(t) = F\,\Pi(t) + \Pi(t)F' + G(t)Q(t)G'(t) \tag{54}$$

The form of K(t,s) and $K_{xy}(t,s)$ confirm the basic assumptions (25) and (32) and also show that the displacement rank is upper-bounded by n ; but we cannot be more explicit than this unless we know the actual form of G(.) and Q(.).

However, suppose G(.) and Q(.) are also constant. Then we note that

$$\ddot{\Pi}(t) = F\,\dot{\Pi}(t) + \dot{\Pi}(t)F' \tag{55a}$$

so that we can write

$$\dot{\Pi}(t) = e^{Ft}\dot{\Pi}(0)e^{F't} \tag{55b}$$

Then we can write, for all t and s,

$$\lrcorner[K(t,s)] = H\,e^{Ft}\dot{\Pi}(0)e^{F's}H' \tag{56}$$

and also

$$\lrcorner[K(t,s)] - K(t,0)K(0,s) = H\ e^{Ft}\dot{P}(0)e^{F's}H' \tag{57}$$

where

$$\dot{P}(0) = \dot{\Pi}(0) - \Pi(0)H'H\,\Pi(0)$$

$$= F\,\Pi(0) + \Pi(0)F' - \Pi(0)H'H\,\Pi(0) + GQG'\ . \tag{58}$$

Now in the Chandrasekhar equations of Section III, the key parameter was the number arising (cf. (13)) in the factorization

$$\dot{P}(0) = L_0 \Lambda L_0 \tag{59}$$

where Λ is an $\alpha \times \alpha$ signature matrix, $\alpha =$ rank $\dot{P}(0)$.
Therefore we can write

$$\lrcorner[K(t,s)] = K(t,0)K(0,s) + L(t)\Lambda L'(s) \tag{60}$$

where

(61) $$L(t) = H e^{Ft} L_0 ,$$

which shows that

$$\alpha = \text{rank } \dot{P}(0) = \text{the displacement rank of}$$

(62) $$K(\cdot,\cdot), \text{ i.e., of the process } H x(.) \ .$$

Thus we have shown the true significance of the somewhat mysteriously occurring parameter α in the Chandrasekhar-equations of Section III ; and more generally, we have properly imbedded a state-space algorithm into a more general input-output recursion.

There remains the question of how to properly imbed the state-space Riccati-equations into an input-output framework. For the completely constant-parameter state-space model, we know that the Riccati equation can be brought into the picture via the relation (cf. the discussion below (13))

(63) $$\dot{P}(t) = L(t)\Lambda L'(t)$$

For processes with completely time-variant state-space models it can be seen from (42) and (47) that $\lrcorner[K(t,s)]$ will not have any convenient form and a displacement rank is difficult to identify.

Various ways of overcoming this difficulty can be considered, e.g., forming higher-order displacement ranks like $\lrcorner^2[K(t,s)]$, but they are only useful in special cases. Some reflection shows that the difficulty is that in our approach we have taken as basic the case of stationary processes and have studied other processes in terms of their distance from stationarity. This will be a good approach for processes that have some connection with time-invariant models, but it may not necessarily be natural for completely time-variant models. In fact, motivated by some results from scattering theory [29], we have discovered [45] that for time-variant state-space models, the dependence of the initial variance $\Pi(t_0)$ on t_0 is the significant feature, and that this can be exploited by studying the rank of the quantity

$$\left(\frac{\partial}{\partial t_0} K(t, s ; t_0)\right) = - K(t, t_0 ; t_0) K(t_0, s ; t_0) \ .$$

However we shall not pursue the details here.

VI. Concluding Remarks

The underlying ideas here allow new computationally efficient solutions for several other problems, both deterministic and stochastic. A discussion of some of these relations can be found in [46] and in [1]. Among more recent examples, we may mention algorithms for online system identification [47] and for the so-called minimal design problems of linear system theory [48-49]. In particular also the new characterizations in terms of distance from stationarity provide new time-invariant implementations for fixed-interval smoothing [50] and Gaussian signal detection [51]. In a return to the radiative-transfer and transport-theory problems, which provided the important Ambartzumian-Chandrasekhar-Sobolev formulas mentioned in Section III, we can adapt our new classifications to obtain a new look at the numerous special functions and integral equations of these theories.

VII. Acknowledgements

As is clear from the list of references, the work described here is part of a continuing research effort with the author's colleague, Professor Martin Morf, and our students.

REFERENCES

[1] T. Kailath, "A View of Three Decades of Linear Filtering Theory", **IEEE Trans. on Information Theory,** vol. IT-20, No. 2, pp. 145-181, March 1974.

[2] H.H. Rosenbrock, **State Space and Multivariable Theory,** New York : J. Wiley, 1970.

[3] V.M. Popov, "Some Properties of Control Systems with Matrix Transfer Functions", in **Lecture Notes in Mathematics,** vol. 144, Berlin : Springer-Verlag, pp. 250-261, 1970.

[4] H.W. Bode and C.E. Shannon, "A Simplified Derivation of Linear Least Square Smoothing and Prediction Theory", **Proc. IRE,** vol. 38, pp. 417-425, April 1950.

[5] L.A. Zadeh and J.R. Ragazzini, "An Extension of Wiener's Theory of Prediction", **J. Appl. Phys.,** vol. 21, pp. 645-655, July 1950.

[6] R.E. Kalman, "A New Approach to Linear Filtering and Prediction Problems", **J. Basic Eng.,** vol. 82, pp. 34-45, March 1960.

[7] R.E. Kalman and R.S. Bucy, "New Results in Linear Filtering and Prediction Theory", **Trans. ASME,** Ser. D.J. Basic Eng., vol. 83, pp. 95-107, Dec. 1961.

[8] R.L. Stratonovich, "Application of the Theory of Markov Processes for Optimum Filtration of Signals", **Radio Eng., Electron. Phys.** (USSR), vol. 1, pp. 1-19, November 1960.

[9] J.L. Casti, R.E. Kalman and V.K. Murthy, "A New Initial-Value Method for On-line Filtering and Estimation", **IEEE Trans. on Information Theory,** vol. IT-18, pp. 515-518, July 1972.

[10] V.A. Ambartsumian, "Diffuse Reflection of Light by a Foggy Medium", **Dokl. Akad. Sci. SSSR,** vol. 38, pp. 229-322, 1943.

[11] S. Chandrasekhar, "On the Radiative Equilibrium of a Stellar Atmosphere, Pt. XXI", **Astrophys. J.,** vol. 106, pp. 152-216, 1947 ; Pt. XXII, **ibid.,** vol. 107, pp. 48-72, 1948.

[11a] S. Chandrasekhar, **Radiative Transfer**, New York : Dover Publications, 1960.

[12] T. Kailath, "Some New Algorithms for Recursive Estimation in Constant Linear Systems", **IEEE Trans. on Information Theory,** vol. IT-19, pp. 750-760, November 1973.

[13] T. Kailath, "Comment on 'A Note on the Use of Chandrasekhar Equations for the Calculation of the Kalman Gain Matrix'," by R.F. Brammer, **IEEE Trans. on Information Theory,** vol. IT-21, pp. 336-337, May 1975.

[14] M. Morf, "Notes on Fast Cholesky Algorithms", Stanford University, 1970 ; see also Ph.D. Dissertation, Stanford University, 1974.

[15] N. Levinson, "The Wiener rms (root-mean-square) Error Criterion in Filter Design and Prediction", **J. Math. Phys.**, vol. 25, pp. 261-278, January 1947.

[16] W. Trench, "An Algorithm for the Inversion of Finite Toeplitz Matrices, **J. SIAM,** vol. 12, pp. 515-522, 1964.

[17] I.C. Gohberg and A.A. Semencul, "On the Inversion of Finite Toeplitz Matrices and their Continuous Analogs", **Math. Issled,** (Russian), No. 2, pp. 201-233, 1972.

[18] I.C. Gohberg and I.A. Fel'dman, "Convolution Equations and Projections Methods for their Solutions", **Translation of Math. Monographs,** vol. 41, Amer. Math. Soc. 1974.

[19] V.V. Sobolev, **A Treatise on Radiative Transfer, Appendices,** Princeton, N.J. : D. Van Nostrand Co., 1963.

[20] L. Ya Geronimus, **Orthogonal Polynomials** (Transl. from the Russian), New York : Consultant's Bureau 1961.

[21] U. Grenander and G. Szegö, **Toeplitz Forms and their Applications,** Berkeley, Ca .: University of Calif. Press, 1958.

[22] T. Kailath, A. Vieira and M. Morf, "Inverses of Toeplitz Operators, Innovations, and Orthogonal Polynomials", **SIAM Review,** Jan. 1978.

[23] P. Whittle, **Prediction and Regulation,** New York : Van Nostrand Reinhold, 1963.

[24] R.A. Wiggins and E.A. Robinson, "Recursive Solution to the Multichannel Filtering Problem", **J. Geophys. Res.,** vol. 70, pp. 1885-1891, April 1965.

[25] M.G. Krein, "The Continuous Analogues of Theorems on Polynomials Orthogonal on the Unit Circle", **Dokl. Akad. Nauk SSSR,** vol. 104, pp. 637-640, 1955.

[26] N.I. Akhiezer, **The Classical Moment Problem,** New York : Hafner Publishing Co., 1965.

[27] L. Ljung, T. Kailath and B. Friedlander, "Scattering Theory and Linear Least Squares Estimation, Part I : Continuous-Time Problems", **Proc. IEEE,** vol. 64, No. 1, pp. 131-138, January 1976.

[28] B. Friedlander, T. Kailath and L. Ljung, "Scattering Theory and Linear Least Squares Estimation, Part II : Discrete-Time Problems", **J. Franklin Institute** vol. 150, January 1976.

[28a] B. Friedlander, T. Kailath and G. Verghese, "Scattering Theory and Linear Least-Squares Estimation", Part III: The Estimates, Proc. 1977 IEEE Conference on Decision and Control, New Orleans, Dec. 1977. See also **IEEE Trans. Automat. Contr.,** vol. AC-25, pp. 794-799, Aug. 1980.

[29] T. Kailath and L. Ljung, "A Scattering Theory Framework for Fast Least-Squares Algorithms", **Proc. Fourth International Multivariate Analysis Symposium,** Dayton, Ohio, June 1975 ; North Holland, Amsterdam, 1977.

[30] T. Kailath, L. Ljung and M. Morf, "A New Approach to the Determination of Fredholm Resolvents of Nondisplacement Kernels", in **Topics in Functional Analysis,** p. 169-184, ed. by I. Gohberg and M. Kac, Acad. Press, 1978.

[31] T. Kailath, L. Ljung and M. Morf, "Recursive Input-Output and State-Space Solutions for Continuous-Time Linear Estimation Problems", **IEEE Trans. Automat. Contr.,** 1982. See also **Proc. IEEE Conf. on Decision and Control,** pp. 182A-185, Florida, Dec. 1976.

[32] B. Friedlander, M. Morf, T. Kailath and L. Ljung, "New Inversion Formulas for Matrices Classified in Terms of Their Distance from Toeplitz Matrices", **J. Lin. Alg. and Applns.,** vol. 27, pp. 31-60, Oct. 1979.

[33] B. Friedlander, T. Kailath, M. Morf and L. Ljung, "Levinson-and Chandraskhar-type Equations for a General Discrete-time Linear Estimation Problem", **IEEE Trans. Automat. Contr.,** pp. 653-659, Aug. 1978.

[34] Y.W. Lee, **Statistical Theory of Communication,** New York : J. Wiley and Sons, 1960.

[35a] V.V. Solodovnikov, **An Introduction to the Statistical Dynamics of Automatic Control Systems,** New York : Dover Publications, 1962 (Russian edition, 1952).

[35b] V.S. Pugachev, **Theory of Random Functions and its Application to Problems of Automatic Control,** Reading, Massachusetts : Addison-Wesley, 1964 (Russian edition, 1959).

[36] H. Laning and R. Battin, **Random Processes in Automatic Control,** New York : McGraw-Hill, 1958.

[37] E.L. Peterson, **Statistical Analysis and Optimization of Systems,** New York : J. Wiley and Sons, 1961.

[38] A.A. Desalu, L.A. Gould and F.C. Schweppe, "Dynamic Estimation of Air Pollution", **IEEE Trans. on Automatic Control,** vol. AC-19, No. 6, pp. 904-910, December 1974.

[39] B.W. Dickinson, T. Kailath and M. Morf, "Canonical Matrix Fraction and State-space Descriptions for Deterministic and Stochastic Linear Systems", **IEEE Trans. on Automatic Control,** vol. AC-19, No. 6, pp. 656-667, December 1974.

[40] R.E. Bellman and G.M. Wing, **An Introduction to Invariant Imbedding,** New York : J. Wiley and Sons, 1974.

[41] M. Morf, and T. Kailath, "Square-root Algorithms for Least-squares Estimation", **IEEE Trans. on Automatic Control,** vol. AC-20, No. 4, pp. 487-497, August 1975.

[42] L.C. Wood and S. Treitel, "Seismic Signal Processing", **Proc. IEEE,** vol. 63, No. 4, pp. 649-661, April 1975.

[43] J. Makhoul, "Linear Prediction: A Tutorial Review", **Proc. IEEE,** vol. 63, No. 4, pp. 561-580, April 1975.

[44] J. Durbin, "The Fitting of Time-series Models", **Rev. Intern. Statist. Inst.,** vol. 28, pp. 213-244, 1960.

[45] B. Friedlander, T. Kailath, M. Morf, "A Modified Definition of Displacement Rank and some Applications", **Proc. 1977 IEEE Conf. on Decision and Control,** pp. 958-961, New Orleans, Dec. 1977.

[46] T. Kailath, "The Innovations Approach to Detection and Estimation Theory", **Proc. IEEE,** vol. 58, No. 5, pp. 680-695, May 1970.

[47] M. Morf, L. Ljung and T. Kailath, "Fast Algorithms for Recursive Identification", **Proc. 1975 IEEE Conf. on Decision and Control,** pp. 916-921, Florida, Dec. 1976.

[48] S. Kung, T. Kailath, M. Morf, "Fast and Stable Algorithms for Minimal Design Problems", **Automatica,** vol. 16, pp. 399-403, July 1980.

[49] S. Kung, M. Morf and T. Kailath, "A Generalized Resultant Matrix for Polynomial Matrices", **Proc. 1976 IEEE Decision and Control Conference,** pp. 892-896, Florida, dec. 1976.

[50] B. Lévy, T. Kailath, L. Ljung and M. Morf, "Fast Time-Invariant Implementations for Linear Least-Squares Smoothing Filters", **IEEE Trans. Automat. Contr.,** AC-24, pp. 770-774, Oct. 1979.

[51] T. Kailath, B. Lévy, L. Ljung and M. Morf, "Fast Time invariant Implementations of Gaussian Signal Detectors", **IEEE Transactions on Information Theory,** vol. IT-24, pp. 469-477, July 1978.

APPENDIX II

ALTERNATIVES IN DISCRETE-TIME RECURSIVE ESTIMATION*

Reprinted by the courtesy of Taylor & Francis Ltd., from Int. J. Control, 1980, vol. 32, no. 2, 311-328.

(*) This work was supported by the Air Force Office of Scientific Research, Air Force Systems Command, under Contract AF44-620-74-C-0068, and in part by the Joint Services Electronics Program under Contract N00014-75-C-0601.

ALTERNATIVES IN DISCRETE-TIME RECURSIVE ESTIMATION

THOMAS KAILATH‡

A summary is presented of several different ways of computing recursive linear least-squares estimates for discrete-time stochastic processes. First focussing on processes with known finite-dimensional models—state-space, ARMA and lumped covariance. Several different algorithms are described—Riccati, square root, Chandrasekhar, fast square root—and also results on convergence and steady-state behaviour. Smoothed estimates and the use of scattering theory to describe their inter-play with filtered estimates are briefly mentioned. Secondly results available for general stationary and non-stationary processes are noted.

1. Introduction

In recent years, several new algorithms and new variants of old algorithms have been developed for linear least-squares estimation of discrete-time processes with or without known finite-dimensional models. In this paper, we have attempted to summarize these new results in an integrated form. Most attention is paid (§ 2) to processes that have known finite-dimensional models, which can be of several types. The most studied have been processes with known state-space models, but one could also consider processes with known ARMA (auto-regressive-moving average) models, and processes with covariance functions given in a certain factored form. In § 3 we more briefly review methods for not necessarily finite-dimensional processes with known covariances.

This paper is an expanded and updated version of a talk presented at an American Geophysical Union Chapman Conference on Applications of Kalman Filtering Theory to Hydrology, Hydraulics and Water Resources, Pittsburgh, Pa., May 1978.

2. Processes with known finite-dimensional models

We start with processes with known state-space models. Thus suppose we have a process $\{y_i, i \geqslant 0\}$ with a known (forwards) state-space model of the form

$$y_i = H_i x_i + v_i, \quad x_{i+1} = F_i x_i + G_i u_i, \quad i \geqslant 0$$

where

$$E u_i u'_j = Q_i \delta_{ij}, \quad E v_i v'_j = R_i \delta_{ij}, \quad E u_i v'_j \equiv 0, \quad E x_0 x'_0 = \Pi_0$$

$$E u_i x'_0 \equiv 0 \equiv E v_i x'_0$$

We assume that $\{F_i, G_i, H_i, Q_i, R_i, \Pi_0\}$ are known matrices with dimensions $n \times n$, $n \times m$, $p \times n$, $m \times m$, $p \times p$ and $n \times n$, respectively; the prime denotes transpose.

2.1. *Kalman filter*

The celebrated Kalman filter (Kalman 1960) can be used to compute

$\hat{y}_{i|i-1} \triangleq$ the linear least-squares estimate (l.l.s.e.) of y_i given $\{y_0, \ldots, y_{i-1}\}$

and

$\hat{x}_{i|i-1} \triangleq$ the l.l.s.e. of x_i given $\{y_0, \ldots, y_{i-1}\}$

The Kalman filter obtains these quantities via the recursions

$$\hat{y}_{i|i-1} = H_i \hat{x}_{i|i-1}, \quad \hat{y}_{0|-1} = 0$$

$$\hat{x}_{i+1|i} = F_i \hat{x}_{i|i-1} + K_i (R_i^{e})^{-1} (y_i - H_i \hat{x}_{i|i-1}), \quad \hat{x}_{0|-1} = 0$$

where $\{K_i, R_i^{e}\}$ are computed via determination of an associated $n \times n$ matrix

$$P_i = E\tilde{x}_i \tilde{x}'_i, \quad \tilde{x}_i \triangleq x_i - \hat{x}_{i|i-1}$$

This is done via the formulas

$$R_i^{e} = R_i + H_i P_i H'_i$$

$$K_i = F_i P_i H'_i$$

where $\{P_i\}$ are obtained via the so-called Riccati difference equation

$$P_{i+1} = F_i P_i F'_i + G_i Q_i G'_i - K_i (R_i^{e})^{-1} K'_i, \quad P_0 = \Pi_0$$

Remark 1

There are numerous variations of the above equations, of which perhaps the most useful is one expressed in terms of time- and measurement-updates : we start with $\hat{x}_{0|-1} = 0$, $P_0 = \Pi_0$ and then successively apply, for $i \geqslant 0$, the equations

Measurement update

$$\hat{x}_{i|i} = \hat{x}_{i|i-1} + P_i H'_i (R_i^{e})^{-1} (y_i - H_i \hat{x}_{i|i-1})$$

$$P_{i|i} = P_i - P_i H'_i (R_i^{e})^{-1} H_i P_i$$

$$(P_{i|i} \triangleq E\tilde{x}_{i|i} \tilde{x}'_{i|i}, \quad \tilde{x}_{i|i} = x_i - \hat{x}_{i|i})$$

Time update

$$\hat{x}_{i+1|i} = F_i \hat{x}_{i|i}$$

$$P_{i+1} = F_i P_{i|i} F'_i + G_i Q_i G'_i$$

Remark 2

There is a vast literature on the Kalman filter. Two recent references emphasizing computational considerations are Bierman (1977) and Bierman and Thornton (1977).

2.2. *Square-root algorithms*

Since P_i is a covariance matrix it must be non-negative-definite, but this property may be lost due to round-off errors in going through the Riccati equation. To avoid this problem, one can instead propagate a 'square-root' of P_i.

Any matrix B such that

$$BB' = A$$

will be called a *square-root* factor of the (necessarily non-negative-definite) matrix A. Square-root factors are not unique because if B is a square-root factor so is BT, where T is an orthogonal matrix ($TT' = 1$). For us, the non-uniqueness is irrelevant because our interest is in the squared quantities†. For its mnemonic value, we shall also write the square-root as $P^{1/2}$,

$$P = P^{1/2}(P^{1/2})' = P^{1/2}P^{T/2}, \quad \text{say}$$

We use the conventions

$$(P^{1/2})^{-1} = P^{-1/2} \quad \text{and} \quad (P^{-1/2})' = P^{-T/2}$$

and also

$$(R^e)^{1/2} = R^{e/2}, \quad (R^e)^{-1/2} = R^{-e/2}$$

We can propagate the $P_i^{1/2}$ as follows: (i) At time i, assume we have $P_i^{1/2}$, and form the array

$$\mathscr{S}_1 = \begin{bmatrix} R_i^{1/2} & H_i P_i^{1/2} & 0 \\ 0 & F_i P_i^{1/2} & G_i Q_i^{1/2} \end{bmatrix}$$

(ii) Apply to it *any* orthogonal transformation Θ such that $\bar{\mathscr{S}}_1$ has zeros as shown

$$\bar{\mathscr{S}}_1 = \begin{bmatrix} X_1 & 0 & 0 \\ X_2 & X_3 & 0 \end{bmatrix} \begin{matrix} p \\ n \end{matrix}$$
$$\qquad\quad p \quad n \quad m$$

(In particular we could require that $\bar{\mathscr{S}}_1$ be lower triangular.) Then the elements appearing in the remaining locations are such that

$$X_3 X'_3 = P_{i+1}$$

so that we can identify X_3 as a square root of P_{i+1}, or

$$X_3 = P_{i+1}^{1/2}$$

It also turns out that

$$X_1 X'_1 = R_i^e, \quad \text{i.e. } X_i = R_i^{e/2}$$

while

$$X_2 X'_1 = K_i$$

† The factors can, of course, be made unique by adding further constraints, e.g. that they be triangular. If one requires that they be symmetric, one obtains a true square-root (whose 'square' equals the original matrix).

Remark 3

The transformation Θ can be carried out in several ways. Explicit knowledge of Θ is not required—it suffices to know the elements of $\mathscr{S}_1$ and that the transformed matrix must have zeros as shown. Householder transformations (cf. Stewart 1973) provide a good way of carrying out this transformation, but other methods are also available (e.g. modified Gram-Schmidt, Givens, unnormalized Givens, etc.—see e.g. Seber (1977), Bierman (1977) and Dobbins (1979)).

By assuming $G_i \equiv 0$, and $F_i = I$, we can obtain the *measurement-update* algorithm

$$\begin{bmatrix} R_i^{1/2} & H_i P_i^{1/2} \\ 0 & P_i^{1/2} \end{bmatrix} \Theta = \begin{bmatrix} R_i^{e/2} & 0 \\ P_i H'_i R_i^{-e/2} & P_{i|i}^{1/2} \end{bmatrix}$$

By assuming $H_i \equiv 0$, we can obtain the *time-update* algorithm

$$\begin{bmatrix} R_i^{1/2} & 0 & 0 \\ 0 & F_i P_{i|i}^{1/2} & G_i Q_i^{1/2} \end{bmatrix} \Theta = \begin{bmatrix} R^{1/2} & 0 & 0 \\ 0 & P_{i+1}^{1/2} & 0 \end{bmatrix}$$

or more simply

$$[F_i P_{i|i}^{1/2} \;\; G_i Q_i^{1/2}]\Theta = [P_{i+1}^{1/2} \;\; 0]$$

We repeat that Θ is used to denote *any* orthogonal transformation such that the resulting array has zeros in the specified locations.

Information forms

In certain problems, especially if Π_0 is very large (e.g. corresponding to large initial uncertainties (which can be incorporated by assuming $\Pi_0 = [\infty]$), it is convenient to propagate $\{P_i^{-1}\}$ or its square-root $\{P_i^{-1/2}\}$. These so-called Information Forms can readily be obtained from the above.

Measurement update

By 'inverting' the relation given above, we see that choosing an orthogonal transformation Θ such that

$$\Theta \begin{bmatrix} P_i^{-1/2} H_i \\ P_i^{-1/2} \end{bmatrix} = \begin{bmatrix} 0 \\ X \end{bmatrix}$$

will yield X such that

$$X'X = P_{i|i}^{-1} \quad \text{or} \quad X = P'^{-1/2}_{i|i}$$

For a *time-update* we should augment the previously given array to give a square matrix, and then invert. Thus we can write

$$\begin{bmatrix} F_i P_{i|i}^{1/2} & G_i Q_i^{1/2} \\ 0 & Q_i^{1/2} \end{bmatrix} \Theta = \begin{bmatrix} P_{i+1}^{1/2} & 0 \\ X_1 & X_2 \end{bmatrix}$$

from which we obtain the rule

$$\Theta\begin{bmatrix} P_{i|i}^{-1/2}\,F_i^{-1} & P_{i|i}^{-1/2}\,F_i^{-1}\,G_i \\ 0 & Q_i^{-1/2} \end{bmatrix} = \begin{bmatrix} P_{i+1}^{-1/2} & 0 \\ X & X \end{bmatrix}$$

Remark 4

There are numerous variations and extensions of the above arrays—see Morf and Kailath (1975), Bierman (1977) and Dobbins (1979). The first reference also explains the origin of the arrays first given. There is reason to believe that square-root implementations are the best way to compute $\{P_i, K_i, R_i^{\mathrm{e}}\}$—see Bierman and Thornton (1977) and Bierman (1977).

2.3. *Constant-parameter models—Chandrasekhar equations*

It is a striking fact that the above Riccati-equation and square-root-array methods are the same whether or not the model parameters $\{F, G, H, Q, R\}$ are constant or vary with time. In fact, we can check that the number of computations (i.e. the number of multiplications and additions) in going from index i to index $i+1$ is $O(n^3)$, whether or not the $\{F, G, H, Q, R\}$ are constant or time-variant. It seems reasonable however that some simplifications should accrue in the constant-parameter case, and in fact we have found how to obtain these (Kailath 1972, 1973, Kailath, Morf and Sidhu 1973, Morf, Sidhu and Kailath 1974, Morf and Kailath 1975, Kailath, Vieira and Morf 1979).

Chandrasekhar equations

Let

$$D \triangleq F\Pi_0 F' + GQG' - F\Pi_0 H'(R+H\Pi_0 H')^{-1} H\Pi_0 F' - \Pi_0$$

and assume that we can represent it (non-uniquely) as

$$D = L_0 M_0 L'_0$$

where L_0 and M_0 are $n\times\alpha$ and $\alpha\times\alpha$ matrices,

$$\alpha = \operatorname{rank} D$$

and

$$M_0 = \operatorname{diag}\{1, 1, \ldots, 1, -1, -1, \ldots, -1\}$$

with as many ± 1's as D has $\pm$ eigenvalues. Then the quantities $\{K_i, R_i^{\mathrm{e}}\}$ appearing in the estimator formula

$$\hat{x}_{i+1|i} = F\hat{x}_{i|i-1} + K_i R_i^{-\mathrm{e}}(y_i - H_i\hat{x}_{i|i-1})$$

can be computed via the equations

$$\begin{array}{c} \\ p \\ n \\ \alpha \\ \\ \end{array}\!\!\begin{array}{c} \begin{array}{cc} p & \alpha \end{array} \\ \begin{bmatrix} R_{i+1}^{\mathrm{e}} & 0 \\ K_{i+1} & L_{i+1} \\ 0 & R_{i+1}^{r} \end{bmatrix} \\ \begin{array}{cc} p & \alpha \end{array} \end{array} = \begin{bmatrix} R_i^{\mathrm{e}} & HL_i \\ K_i & L_i \\ L'_i H' & R_i^{r} \end{bmatrix} \begin{bmatrix} I & -(R_i^{\mathrm{e}})^{-1} HL_i \\ -(R_i^{r})^{-1} L'_i H' & I \end{bmatrix}$$

where the initial values are

$$\{R_0^e = R + H\Pi_0 H\,,\quad K_0 = F\Pi_0 H' \quad \text{and} \quad R_0^r = -M_0^{-1}\}$$

Remark 5

The number of computations required to go from index i to index $i+1$ can be seen to be $O(n^2\alpha)$ as compared to $O(n^3)$ if the Riccati equation is used. There will be a computational advantage if it happens (or can be arranged) that $\alpha < n$.

Example 1

If $\Pi_0 = 0$, then $D = GQG'$. If we assume G to have full rank (and $m < n$) then $\alpha = m$ and we can take

$$L_0 = GQ^{1/2}, \quad M_0 = I$$

Example 2. *Stationary processes*

If F is stable and Π_0 is such that $F\Pi_0 F' + GQG' = \Pi_0$, then it can be shown that the processes $\{x_i, y_i\}$ are stationary. In this case,

$$D = -F\Pi_0 H'(R + H\Pi_0 H')^{-1} H\Pi_0 F'$$

If we assume H to be full rank (and $p < n$), then $\alpha = p$, and we can take

$$L_0 = F\Pi_0 H'(R + H\Pi_0 H')^{-T/2}, \quad M_0 = -I$$

Remark 6

The result of Example 2, namely for the special case of stationary processes, was also obtained by Lindquist (1974) by an approach that provides no guidance for the non-stationary case.

Remark 7

If α does not happen to be low, it seems reasonable that one should try to vary $\{F, G, H, Q, R, \Pi_0\}$ slightly until we get a low value of α. If this is for some reason not permissible then one could deliberately use an incorrect value of Π_0 and then correct for this error later—in certain problems this procedure can still be computationally useful (see Ljung and Kailath 1977).

Remark 8

One might wonder about the physical significance of the parameter α. It turns out that α is a measure of how 'close to stationary' the process y is—see Kailath (1975), Kailath, Ljung and Morf (1978), Friedlander, Kailath, Morf and Ljung (1978) and Kailath, Kung and Morf (1979 a, b).

Remark 9

The *Riccati variable* does not enter the above equations. However if desired it can be computed as

$$P_{i+1} = \Pi_0 - \sum_0^i L_j (R_j^e)^{-1} L'_j$$

Remark 10

The equations for $\{K_i, L_i\}$ were called *Chandrasekhar equations* because they are the discrete-time form of (a generalization of) certain non-linear differential equations first introduced by Chadrasekhar in 1947 to solve the finite-interval Wiener–Hopf equation for a special class of stationary (or difference or convolution) kernels arising in radiative transfer theory (see Kailath 1972, 1973, 1974).

2.4. *Constant parameter models—fast square-root algorithms*

There exist square-root versions of the Chandrasekhar equations. We shall consider first the special case $\Pi_0=0$. In this case it turns out that

$$P_{i+1}-P_i \geqslant 0, \quad i \geqslant 0$$

and therefore we can factor it as

$$P_{i+1}-P_i=\bar{L}_i\bar{L}'_i$$

Also define

$$\bar{K}_i=K_i(R_i^{\mathrm{e}})^{-\mathrm{T}/2}$$

$$L_0=GQ^{1/2}, \quad \bar{K}=0, \quad R_0^{\mathrm{e}}=R$$

Then we have the square-root arrays

$$\begin{array}{c} \\ p \\ n \end{array}\begin{array}{c} \begin{array}{cc} p & m \end{array} \\ \begin{bmatrix} R_i^{\mathrm{e}/2} & H\bar{L}_i \\ \bar{K} & F\bar{L}_i \end{bmatrix} \end{array} \mathscr{T} = \begin{array}{c} \begin{array}{cc} p & m \end{array} \\ \begin{bmatrix} R_{i+1}{}^{\mathrm{e}/2} & 0 \\ \bar{K}_{i+1} & \bar{L}_{i+1} \end{bmatrix} \end{array}\begin{array}{c} \\ p \\ n \end{array}$$

The Riccati variable can be computed if desired as

$$P_{i+1}=\sum_{0}^{i} \bar{L}_j\bar{L}'_j$$

Remark 11

When compared with the general (i.e. for possibly time-variant F, G, ...) arrays of § 2.2, we have a reduction in storage from $O((n+p+m)^2)$ to $O((n+p)(p+m))$ and a reduction in computations from $O((n+p+m)^3)$ to $O((n+p)(p+m)^2)$.

General case $\Pi_0 \neq 0$

In this case, as may be expected from the Chadrasekhar equations of § 2.3, we shall have to use J-orthogonal transformations Θ such that

$$\Theta J \Theta'=J$$

where J is a signature matrix of the form diag $\{1, 1, 1, \ldots, -1, -1, \ldots, -1\}$. For details we refer to Kailath, Vieira and Morf (1979), where the general fast square-root arrays are deduced from the general Chandrasekhar equation.

2.5. *Constant-parameter models—asymptotic properties*

In order to ensure that the effect of round-off errors will not accumulate, one should examine whether the solution of the Riccati equation converges to the same steady-state value, irrespective of the choice of initial condition. This has been a widely studied problem; perhaps the best result to date has been found by Kailath and Ljung (1976) (see also Ljung, Kailath and Friedlander (1976, Sec. III.D)).

The main result is the following: if $\{F, G\}$ is stabilizable and $\{H, F\}$ is detectable, then the solution of the Riccati equation converges to a constant solution $\bar{P}$ as $i \to \infty$ for all symmetric initial conditions Π_0 such that (this was mistyped in eqn. (22) of Kailath and Ljung (1976))

$$x'(\bar{P}^a \Pi_0 \bar{P}^a + \bar{P}^a)x > 0$$

for all x such that $\bar{P}^a x \neq 0$, where

$$\bar{P}^a = \lim_{i \to \infty} P_0^a(i)$$

and $P_0^a(i)$ obeys the 'adjoint' Riccati equation

$$P_0^a(i+1) = FP_0^a(i)F' + GQG' - FP_0^a(i)H'(R + HP_0^a(i)H')^{-1} \times HP_0^a(i)F', \quad P_0^a(0) = 0$$

Remark

$\bar{P}^a$ always exists under the assumptions on $\{H, F, G\}$ and is the unique non-negative-definite solution of the adjoint steady-state equation obtained from the above equation by setting $P_0^a(i+1) = P_0^a(i)$.

Special cases

(i) If $\Pi_0 > 0$, the condition is trivially met—this is the result of Wonham (for $\{H, F\}$ observable) and Kucera.

(ii) If $\{H, F\}$ is observable, then $\bar{P}^a$ is invertible and the condition reduces to

$$\Pi_0 > -(\bar{P}^a)^{-1}$$

It can be shown that when $\{H, F\}$ is observable, we have $\bar{P}^a = \bar{P}$, the *infimum* over all solutions to the steady-state Riccati equation (A.R.E.)

$$\bar{P} = F\bar{P}F' + GQG' - F\bar{P}H'(R + H\bar{P}H')^{-1}H\bar{P}F'$$

The convergence result for this case was first obtained by Willems (1971) and by Rodriguez-Canabal (1973).

(iii) An example not covered by any earlier results is

$$F = \begin{bmatrix} -1 & 0 \\ 0 & 1 \end{bmatrix}, \quad G = \begin{bmatrix} 2 & 0 \\ 0 & 3 \end{bmatrix}, \quad H = [0 \ \ 1]$$

for which $\bar{P}^a$ and $\bar{P}$ turn out to be

$$\bar{P}^a = \begin{bmatrix} 0 & 0 \\ 1 & 1 \end{bmatrix}, \quad \bar{P} = \begin{bmatrix} 1 & 0 \\ 0 & -1 \end{bmatrix}$$

Our condition then assures convergence for all initial conditions such that

$$\Pi_0 = \begin{bmatrix} \Pi_{11} & \Pi_{12} \\ \Pi_{21} & \Pi_{22} \end{bmatrix}, \quad \{\Pi_{11}, \Pi_{12}, \Pi_{13}\} \text{ arbitrary, } \Pi_{21} > -1$$

while the condition $\Pi > \bar{P}$ is much more restrictive.

Apart from the stronger results, we would also emphasize the simplicity and directness of the proofs in Kailath and Ljung (1976), as compared to those in the earlier literature (Willems 1971, Rodriguez-Canabal 1973).

Solution of the A.R.E.

Assuming that $P(i) \to \bar{P}$ as $i \to \infty$, it is of interest to actually determine $\bar{P}$, because the resulting steady-state Kalman filter will be time-invariant and easier to implement. In fact, the steady-state Kalman filter is often used as a simple approximate filter even when $i < \infty$. It is not difficult to estimate the consequent penalty incurred in the mean-square error.

One method of finding P is to increase i until $P(i)$ does not change 'much'—however it may take a long time for this to happen and also the accuracy may be poor if there is a big spread in the values of the eigenvalues of $(F - \bar{K}H)$. Letting $i \to \infty$ in the Chandrasekhar equations seems to work better (see the numerical examples in Kailath (1973) and Casti and Kirschner (1976)).

Another method is to tackle the A.R.E. directly by finding the eigenvalues and eigenvectors of a matrix derived from the $2n \times 2n$ hamiltonian matrix

$$\begin{bmatrix} F & GQG' \\ -H'R^{-1}H & F' \end{bmatrix}$$

This method was suggested by MacFarlane (1963) and Potter (1966) and developed into a satisfactory numerical algorithm by Fath (1969) and by Bryson and Hall (1971). This eigenvalue–eigenvector method works well when the eigenvalues of the hamiltonian matrix are distinct.

Recently we have developed (see Morf, Dobbins *et al.* 1979) a different, so-called 'square-root-doubling' approach to the direct determination of the limiting value of the Riccati equation by successively computing the values

$$P^{1/2}(1),\ P^{1/2}(2),\ P^{1/2}(4),\ \ldots,\ P^{1/2}(2^i)$$

The relevant arrays are

$$\begin{bmatrix} I & W_i^{T/2} P_i^{1/2} & 0 \\ 0 & \Phi_{*i} P_i^{1/2} & P_i^{1/2} \end{bmatrix} \mathscr{T}_1 = \begin{bmatrix} R_{*i}^{1/2} & 0 & 0 \\ K_{*1} & P_{2i}^{1/2} & 0 \end{bmatrix}$$

$$\mathscr{T}_2 \begin{bmatrix} R_{*i}^{1/2}\, W_i^{T/2}\, \Phi_{*i} \\ W_i^{T/2} \end{bmatrix} = \begin{bmatrix} 0 \\ W_{2i}^{T/2} \end{bmatrix}$$

$$\Phi_{*2i} = (I - K_{*i} R_{*i}^{-1/2} W_i^{T/2}) \Phi_{*i}$$

where $\{R_{*i}, K_{*i}\}$ are certain auxiliary quantities and

$$\Phi_{*1}=F, \quad P_1^{1/2}=GQ^{1/2}, \quad W_1^{\mathrm{T}/2}=R^{-\mathrm{T}/2}H$$

In this way, in a few steps one reaches a very large value of the index i (e.g. in 20 steps one essentially has $P^{1/2}(10^6)$) and therefore round-off errors do not accumulate as fast as in a direct solution of the Riccati equation. Also the method is indifferent to the presence of repeated eigenvalues in the hamiltonian matrix or to the possible singularity of F. Further numerical comparisons and physical interpretations are pursued in the thesis of Newkirk (1979).

We may note that the steady-state filter (when F is stable) can be found directly by the celebrated spectral-factorization technique of Wiener and Hopf, which is based on finding the poles and zeros of the power-spectral density function $S_y(z)$ (cf. § 2.8). For scalar processes, this can often be simpler than solving via Riccati or Chandrasekhar equations.

2.6. *Smoothed estimates and scattering theory*

There are problems in which causal estimates $\{\hat{x}_{i|i-1}$ or $\hat{x}_{i|i}\}$ are not essential but we can use future data (of course with a delay or off-line) to get better estimates. Such estimates, e.g. $\hat{x}_{i|j}$, $j>i$, are often known as smoothed estimates. While in principle the smoothed estimates are completely determined by the innovations (and hence by the one-step predicted estimates), there are in fact several different ways of computing the smoothed estimates—e.g. via the innovations (Kailath–Frost), via the filtered estimates (Rauch, Tung–Striebel), via forward and backward causal filters (Mayne–Fraser) (which are closely related to the so-called partitioning formulas of Lainiotis (1976)). It turns out that a convenient way of organizing and deriving these formulas (and various others) is provided by the use of a certain transmission-line scattering theory framework (see Ljung and Kailath 1975, Verghese, Friedlander and Kailath 1977). We shall not elaborate on this here except to note that the scattering-theory framework is a natural one in studying problems of signal transmission and reflection in layered-earth media (Robinson 1967, Claerbout 1975).

In such applications, real-time or causal estimation may not be as significant as it is in control problems, and smoothed estimates can be used to obtain lower mean-square-error. In this connection, it is important to note that the solution of the smoothing (but not the causal filtering) problem can be implemented via *time-invariant* causal and anti-causal) filters, which fact can be of value in implementation and, for example, in enabling the use of F.F.T. techniques—this new approach to the smoothing problem is described in Lévy *et al.* (1979 a), (see also Kailath, Lévy *et al.* (1978) for applications to signal detection and also Lévy *et al.* (1979 b)).

2.7. *A.R.M.A. models*

Suppose we have process models of the form

$$y_i+\sum_{j=1}^{n} a_j(i)y_{i-j}=\sum_{j=0}^{m} b_j(i)u_{i-j}, \quad i\geqslant m, \; m\leqslant n$$

with given initial conditions $\{y_0, \ldots, y_{m-1}\}$ and with

$$Eu_i u'_j = \delta_{ij}, \quad Eu_i y'_j = 0, \quad i \geqslant j$$

and the $\{a_j(\cdot), b_j(\cdot)\}$ are known scalar functions (the case of matrix $\{a(\cdot), b(\cdot)\}$ can also be studied). We can convert such models to state-space form but we shall see that it can be simpler (when $m < n$) to proceed directly. Let us first rewrite the model as

$$y_i + \sum_{j=1}^{n} a_j(i) y_{i-j} = w_i, \quad i \geqslant 0$$

where we define

$$w_i = \sum_{j=0}^{m} b_j(i) u_{i-j}, \quad i \geqslant n$$

and

$$w_i \triangleq y_i - \hat{y}_{i|[i-1,0]} \quad \text{for } i = 0, \ldots, n-1$$

To find the one-step predictions $\{\hat{y}_{i|i-1}\}$ for such models note that

$$\hat{y}_{i|i-1} + \sum_{j=1}^{n} a_j(i) y_{i-j} = \hat{w}_{i|i-1}$$

where

$$\hat{w}_{i|i-1} = \text{the l.l.s.e. of } w_i \text{ given } \{y_0, \ldots, y_{i-1}\}$$

But we note from the specification w_i that we have a causal and causally invertible mapping between $\{y_0, \ldots, y_i\}$ and $\{w_0, \ldots, w_i\}$, $i \geqslant 0$. Therefore we can regard w_i as

$$\hat{w}_{i|i-1} = \text{the l.l.s.e. of } w_i \text{ given } \{w_0, \ldots, w_{i-1}\}$$

The point is that from the above, we have

$$y_i - \hat{y}_{i|i-1} = w_i - \hat{w}_{i|i-1}$$

so the problem of finding the innovations of $\{y_i\}$ is the same as that of finding the innovations of $\{w_i\}$. If $m \ll n$, then $\{w_i\}$ is a simpler (so-called *moving-average*) process and it can be easier to work with it than with the process $\{y_i\}$. But once the innovations have been found for the simpler process $\{w_i\}$ (by any of the methods of §§ 2.1 to 2.4), then we can translate the results back to apply to the actually observed process $\{y_i\}$. For more details see Rissanen and Barbosa (1969), Gupta and Kailath (1978), Morf, Kailath and Dickinson (1974), Dickinson, Kailath and Morf (1974) and also Morf (1974) and Sidhu 1975).

2.8. *Lumped covariance models*

Suppose we do not have a state-space or A.R.M.A. process model, but only have measurements of the covariance function or more often (under the assumption of stationarity) of the power-spectral density of the signal and the noise processes.

Suppose in the latter case that we fit a rational transfer function to the measured power spectral density of the signal plus (uncorrelated) noise process.

Equivalently suppose we fit a triple $\{M, F, N\}$ (by standard state-space minimal realization methods) such that

$$S_y(z) = I + M(zI - F)^{-1}N + N'(z^{-1}I - F')M'$$

or in the time-domain

$$R_y(k-l) = \begin{cases} MF^{k-l}N, & k>l \\ I, & k=l \\ N'F'^{(l-k)}M', & k<l \end{cases}$$

One approach is to find a state-space model for the signal process to match the given signal covariance and then to apply the Kalman filter. This can be done but takes a substantial effort (see, e.g. Anderson and Moylan 1974). It is possible to work directly with the $\{M, F, N\}$ quantities to obtain either Riccati or Chandrasekhar or square-root algorithms (see Kailath and Geesey 1971, Morf, Sidhu and Kailath 1974, Morf and Kailath 1975). We shall not give the details here.

3. Processes without known finite-dimensional structure

There are many applications in which state-space or A.R.M.A. models are not readily at hand or are not easy to construct. In such cases we cannot of course obtain estimates as conveniently but there is still a substantial amount that can be done.

Thus suppose we have observations of a random process $\{y_0, y_1, \ldots, y_i, \ldots\}$ with known covariances

$$R_{ij} = Ey_iy'_j$$

and that we wish to estimate a related random process $\{x_i\}$ or to predict the future values of the process $\{y_i\}$ itself. Solution of the latter problem is really the key to the more general problem, and therefore we shall say a few words about it here.

Let us write

$$\hat{y}_{N|N-1} = -A_{N,1}y_{N-1} \cdots - A_{N,N}y_0$$

where the $\{A_{N,i}\}$ are to be chosen to minimize the mean-square-error

$$E[y_N - \hat{y}_{N|N-1}]^2$$

For this to be true we must have

$$e_N \triangleq (y_N - \hat{y}_{N|N-1}) \perp \{y_0, \ldots, y_{N-1}\}$$

where $a \perp b$ signifies $Eab = 0$. If we also define

$$R_N^{e} = Ee_Ne'_N = Ee_Ny'_N$$

then it is easy to see that the above conditions lead to the equations

$$A_NR_N = [0 \ \ldots \ R_N^{e}]$$

where

$$A_N \triangleq [A_{N,N} \; \cdots \; A_{N,1} \quad I]$$

$$R_N \triangleq [E y_i y'_j]_{i,j=0}{}^N$$

These linear equations can of course be solved in standard ways, but at the cost of $O(N^3)$ operations (multiplications and additions). To reduce this burden, and also to have convenient ways of updating the solutions, A_N, we need to impose more structure on the process. We briefly note what is known at present.

3.1. *Stationary processes*

When the process is stationary, $R_{ij} = R_{i-j}$ and R_N becomes a Toeplitz matrix. In this case the now famous L.W.W.R. (Levinson–Whittle–Wiggins–Robinson) algorithm (see Levinson 1974, Whittle1963, Wiggins and Robinson 1965) can be used to compute $\{A_N\}$ recursively with $O(N^2)$ computations, an *order-of-magnitude* reduction. The algorithm proceeds as follows

$$\begin{bmatrix} R_{N+1}{}^{\mathrm{e}} & 0 \\ \hdashline A'_{N+1} & B'_{N+1} \\ \hdashline 0 & R_{N+1}{}^{\mathrm{r}} \end{bmatrix} \doteq \begin{bmatrix} R_N{}^{\mathrm{e}} & \Delta_N \\ \hdashline 0 & B'_N \\ A'_N & 0 \\ \hdashline \Delta'_N & R_N{}^{\mathrm{r}} \end{bmatrix} \Theta_{N+1}$$

where

$$\Theta_{N+1} \triangleq \begin{bmatrix} I & -R_N{}^{-\mathrm{e}}\,\Delta_{N+1} \\ -R_N{}^{-\mathrm{r}}\,\Delta'_{N+1} & I \end{bmatrix}$$

$$\Delta_{N+1} = \sum_{i=0}^{N} A_{N,N-i} R_i, \quad N = 1, \ldots,$$

and the initial conditions are

$$R_0{}^{\mathrm{e}} = R_0 = R_0{}^{\mathrm{r}}, \quad A_0 = I = B_0$$

Again the quantities $\{B_N, \Delta_N, R_N{}^{\mathrm{r}}\}$ are auxiliary quantities that do have physical significance. Here however we wish to note that the Levinson recursions are of the same form as the Chandrasekhar equations in § 2.3. In fact, it can be shown that if the *stationary* process has a known state-space model, then the Levinson arrays reduce to the Chandrasekhar ones (with $A'_{N+1} \to K_{N+1}$, $[0 \;\; A'_N] \to K_N$, $B'_{N+1} \to L_{N+1}$, $[B'_N \;\; 0] \to FL_N$, $\Delta_N \to HL$). Moreover the matrix Θ_N can be interpreted as the 'propagation' matrix of an elementary 'lossless' section of some layered (transmission-line) medium. Among other things, this interpretation can be used to obtain normalized 'array forms' of the L.W.W.R. algorithm (Kailath, Vieira and Morf 1978) and to make useful connections with the Darlington synthesis methods of network theory and with certain interpolation problems (cf. Schur, Nevanlinna–Pic, etc.) of classical function theory (Dewilde, Vieira and Kailath 1978).

A symmetrized form of the above L.W.W.R. algorithm is useful here (see Morf, Vieira and Kailath 1978).

3.2. *Processes close to stationary*

When $\{y_i\}$ is not stationary, R_N is no longer Toeplitz. But in recent work we have shown that we can define an integer α that provides an index of how 'close' R_N is to being Toeplitz (Kailath, Ljung and Morf 1978, Kailath, Kung and Morf 1979 a, b). This index, say α, can be defined as unity for Toeplitz matrices, and has the significance that when R_N has index α it takes $O(N^2\alpha)$ computations to compute $\{A_N\}$. Moreover, this can be done by a generalized form of the L.W.W.R. algorithm. It is striking in fact that the only changes are that B is now an $N \times \alpha$ block matrix (rather than just $N \times 1$) and that the formula for Δ_N has to be slightly modified. Moreover, if it happens that the process has a constant-parameter state-space model of order n, then $\alpha \leqslant n$, and if the state-space model is known then the generalized Levinson algorithm can be reduced to the Chandrasekhar equations of § 2.3. These results are discussed in detail in Friedlander *et al.* (1978, 1979).

An important situation in which non-stationarity is encountered is when very little *a priori* statistical information is available about the stochastic processes involved. Often we may only have a single record of observations, $\{y_0, \ldots, y_N\}$, of a process whose future values we would like to estimate, with no information about the means and covariances of the processes. It is reasonable here to attempt to use the classical deterministic least-squares criterion: for example, choose $\{a_1, \ldots, a_p\}$ so that

$$\sum_t e_{p,t}^2 = \text{minimum}$$

where

$$e_{p,t} = y_t + a_1 y_{t-1} + \ldots + a_p y_{t-p}$$

is the pth order residual at time t. However, we shall have to make some assumption about the 'missing data', $\{y_{-p}, \ldots, y_{-1}\}$ and $\{y_{N+1}, \ldots, y_{N+p}\}$, that will be needed to form the residuals $\{e_p, \ldots, e_{p-1}\}$ and $\{e_{p-N+1}, \ldots, e_{N+p}\}$, should we decide to use some or all of them. Various assumptions can be made (see, e.g. Makhoul 1975, Morf–Lee *et al.* 1977, Morf, Vieira and Lee 1977), which generally lead to problems requiring the inversion of non-Toeplitz matrices. However, the special assumptions involved allow us to identify the index α of 'non-toeplitzness' and thus to use the generalized Levinson algorithms mentioned above (see Lee 1980, Lev-Ari 1980).

3.3. *Ladder-form implementations*

The previously mentioned connection of the Θ-matrices to transmission-line sections leads to a useful so-called 'ladder-form' implementation of the above algorithms. Let us consider for simplicity the case of scalar, stationary processes for which it is easy to see that we can write

$$\Theta_N = \begin{bmatrix} 1 & -\rho_N \\ -\rho_N & 1 \end{bmatrix}, \quad \rho_N = -\Delta_N / R_{N-1}^e$$

Then the innovation e_N can be computed by feeding $\{y_0, \ldots, y_N\}$ into the 'feedforward' ladder structure. If an additional data point is to be handled, were merely add on another section Θ_{N+1}, without disturbing the previous sections.

This is in sharp contrast to implementation via transversal filters (or tapped delay lines) of the impulse response $A_N(z^{-1})$; in this case, handling of an additional observation would require not only an additional tap but also a change of every previous tap gain so as to realize an impulse response $A_{N+1}(z^{-1})$. Thus one needs a time-variant growing memory tapped delay line filter, as opposed to the time-invariant growing memory ladder filter.

The growing memory feature is unavoidable unless the process is truly autoregressive. If the process is known to have finite-dimensional structure, then a balance has to be struck between trying to identify this structure or using a suitably long ladder filter. We should emphasize also that the finite-dimensional fixed-memory filters described in § 2 are generally time-variant, while the ladder filters are time-invariant, a point that is important in hardware (especially integrated circuit) implementation.

Further studies of these important ladder-form realizations can be found in the theses of D. Lee (1980) and H. Lev–Ari (1980).

4. Concluding remarks

In this paper we have reviewed the various algorithms now available for linear least-squares estimation of stochastic processes. Processes with additional finite-dimensional structure (given via known state-space *or* known A.R.M.A. models *or* via known lumped covariance descriptions) have been the most studied in the last two decades, partly because several fixed-memory (but generally time-variant) recursive estimators can be specified for them. We have organized and described these algorithms in § 2.

In signal processing applications, attention has been turning again to processes for which the assumption of finite-dimensionality is not helpful—either because finite-dimensional models are inadequate or because they are too hard to determine from the available information. In § 3 we have briefly described the generalized Levinson algorithms and growing memory but time invariant ladder form implementations that can be used for such processes. The author believes that these results will see increasing application and hardware realization in a variety of signal processing problems.

References

Anderson, B. D. O., and Moylan, P. J., Spectral factorization of a finite-dimensional nonstationary matrix covariance, *I.E.E.E. Trans. autom. Control*, **19**(6), 680–693.

Bierman, G. J., 1977, *Factorization Methods for Discrete Sequential Estimation* (New York: Academic Press).

Bierman, G. J., and Thornton, C. L., 1977, Numerical comparison of Kalman filter algorithms: orbit determination case study. *Automatica*, **13**, 23–35.

Bryson, A. E., and Hall, W., 1971, Optimal control synthesis by eigenvector decomposition. Tech. Rept. 436, Dept. of Aero & Astro, Stanford University, Stanford, CA.

Casti, J., and Kirschner, O., 1976, Numerical experiments in linear control theory using generalized x, y equations. *I.E.E.E. Trans. autom. Control*, **21** (5), 792–795.

CLAERBOUT, J., 1975, *Fundamentals of Geophysical Data Processing* (New York : McGraw-Hill).

DEWILDE, P., VIEIRA, A., and KAILATH, T., 1978, On a generalized Szego–Levinson realization algorithm for optimal linear predictions based on a network synthesis approach. *I.E.E.E. Trans. Circuits Systems*, **25** (9), 663–675.

DICKINSON, B., KAILATH, T., and MORF, M., 1974, Canonical matrix fraction and state-space descriptions for deterministic and stochastic linear systems. *I.E.E.E. Trans. autom. Control*, Special Issue on System Identification and Time Series Analysis, **19** (6), 656–667.

DOBBINS, J., 1979, Covariance factorization techniques for least squares estimation. Ph.D. Dissertation (Dept. of Elect. Engng, Stanford University, Stanford, CA).

FATH, A. F., 1969, Computational aspects of the linear optimal regulator problem. *I.E.E.E. Trans. autom. Control*, **14** (5), 547–550.

FRIEDLANDER, B., KAILATH, T., MORF, M., and LJUNG, L., 1978, Extended Levinson and Chandrasekhar equations for general discrete-time linear estimation problems. *I.E.E.E. Trans. autom. Control*, **23** (4), 653–659.

FRIEDLANDER, B., MORF, M., KAILATH, T., and LJUNG, L., 1979, New inversion formulas for matrices classified in terms of their distance from Toeplitz matrices. *Lin. Alg. Applns*, **27**, 31–60.

GUPTA, N., and KAILATH, T., 1978, Corrections and extensions to innovations, Pt. VII : Some applications of vector A.R.M.A. models. *I.E.E.E. Trans. autom. Control*, **23**, 511–512.

KAILATH, T., 1972, Some Chandrasekhar-type algorithms for quadratic regulators. *Proc. I.E.E.E. Conf. Decision Control*, New Orleans, LA, pp. 219–223.

KAILATH, T., 1973, Some new algorithms for recursive estimation in constant linear systems. *I.E.E.E. Trans. Inform. Thy*, **19** (6), 750–760.

KAILATH, T., 1974, A view of three decades of linear filtering theory. *I.E.E.E. Trans. Inform. Thy*, **20** (2), 145–181.

KAILATH, T., 1975, Some new results and insights in linear least-squares estimation theory. *Proc.* 1975 *I.E.E.E.–USSR Joint Workshop on Inform. Thy*, Moscow, USSR, pp. 97–104. Reprinted with corrections in Kailath (1978).

KAILATH, T., 1978, *Lectures in Linear Least Squares Estimation* (New York : Springer-Verlag).

KAILATH, T., and GEESEY, R., 1971, An innovations approach to least squares estimation, Pt. IV : Recursive estimation given the covariance functions. *I.E.E.E. Trans. autom. Control*, **16** (6), 720–727.

KAILATH, T., KUNG, S.-Y., and MORF, M., 1979 a, Displacement ranks of matrices and linear equations. *J. Math. Anal. Applns*, **68** (2), 395–407.

KAILATH, T., KUNG, S.-Y., and MORF, M., 1979 b, Displacement ranks of a matrix. *Bulletin Amer. Math. Soc.*, **1** (5), 769–773.

KAILATH, T., LÉVY, B., LJUNG, L., and MORF, M., 1978, Fast time-invariant implementations of gaussian signal detectors. *I.E.E.E. Trans. Inform. Thy*, **24** (4), 469–477.

KAILATH, T., and LJUNG, L., 1976, Asymptotic behaviour of constant-coefficient Riccati differential equations. *I.E.E.E. Trans. autom. Control*, **21** (3), 385–388.

KAILATH, T., LJUNG, L., and MORF, M., 1978, Generalized Krein–Levinson equations for efficient calculation of Fredholm resolvents of nondisplacement kernels. In *Topics in Functional Analysis, Dedicated to M. G. Krein on the Occasion of His* 70*th Birthday*, edited by I. Gohberg and M. Kac (New York : Academic Press).

KAILATH, T., MORF, M., and SIDHU, G. S., 1973, Some new algorithms for recursive estimation in constant linear discrete-time systems. *Proc. Seventh Princeton Conf. Inform. Sci. Systs*, Princeton, N.J., pp. 344–352.

KAILATH, T., VIEIRA, A., and MORF, M., 1978, Orthogonal transformation (square-root) implementations of the generalized Chandrasekhar and generalized Levinson algorithms, *Proc.* 1978 *IRIA Conf. on Syst., Optimization & Analysis*, edited by A. Bensoussan and J. Lions (New York : Springer-Verlag), pp. 81–91.

Kalman, R. E., 1960, A new approach to linear filtering and prediction problems. *J. Basic Engng*, **82,** 34–45.

Lainiotis, D. G., 1974, Partitioned estimation algorithms, II : Linear estimation. *Inform. Sci.*, **7,** 317–340.

Lee, D., 1980, Ph.D. Dissertation (Dept. of Elect. Engng, Stanford University, Stanford, CA).

Lev-Ari, H., 1981, Ph.D. Dissertation (Dept. of Elect. Engng, Stanford University, Stanford, CA).

Levinson, N., 1974, The Wiener r.m.s. (root-mean-square) error criterion in filter design and prediction. *J. Math. Phys.*, **25,** 261–278.

Lévy, B., Kailath, T., Ljung, L., and Morf, M., 1979 a, Fast time-invariant implementations of linear least-squares smoothing filters. *I.E.E.E. Trans. autom. Control*, **24** (5), 770–774.

Lévy, B., Kailath, T., Ljung, L., and Morf, M., 1979 b, The factorization and representation of operators in the algebra generated by Toeplitz operators. *SIAM J. Appl. Math.*, **37** (3), 467–484.

Lindquist, A., 1974, A new algorithm for optimal filtering of discrete-time stationary processes. *SIAM J. Control*, **12** (4), 736–747.

Ljung, L., and Kailath, T., 1975, A scattering theory framework for fast least-squares algorithms. *Fourth Int'l. Symp. on Multivariate Analysis*, Dayton, OH. Published in *Multivariate Analysis—IV*, edited by P. R. Krishnaiah (1977) (Amsterdam : North-Holland), pp. 387–406.

Ljung, L., and Kailath, T., 1976, A unified approach to smoothing formulas. *Automatica*, **12** (2), 147–157.

Ljung, L., and Kailath, T., 1977, Efficient change of initial conditions, dual Chandrasekhar equations and some applications. *I.E.E.E. Trans. autom. Control*, **22** (3), 443–447.

Ljung, L., Kailath, T., and Friedlander, B., 1976, Scattering theory and linear least squares estimation, Pt. I : Continuous-time problems. *Proc. Inst. elect. electron. Engrs*, **64** (1), 131–139.

MacFarlane, A. G. J., 1963, An eigenvector solution of the optimal linear regulator. *J. Electron. Control*, **14,** 643–654.

Makhoul, J., 1975, Linear prediction : a tutorial review. *Proc. Inst. elect. electron. Engrs*, **63,** 561–580.

Morf, M., 1974, Fast algorithms for multivariable systems. Ph.D. Dissertation (Dept. of Elect. Engng, Stanford University, Stanford, CA).

Morf, M., Dobbins, J., Friedlander, B., and Kailath, T., 1979, Square-root algorithms for parallel processing in optimal estimation. *Automatica*, **15,** 299–306.

Morf, M., Kailath, T., and Dickinson, B., 1974, General speech models and linear estimation theory. Presented at *I.E.E.E. Speech Processing Conf.*, Carnegie-Mellon University. Appears in *Speech Recognition*, edited by R. Reddy (1975) (New York : Academic Press), pp. 157–182.

Morf, M., and Kailath, T., 1975, Square-root algorithms for least-linear squares estimation. *I.E.E.E. Trans. autom. Control*, **20** (4), 487–497.

Morf, M., Lee, D., Nickolls, J., and Vieira, A., 1977, A classification of algorithms for A.R.M.A. models and ladder realizations. *Proc.* 1977 *I.E.E.E. Int'l. Conf. on Acoustics, Speech Signal Processing*, Hartford, CT, pp. 13–19.

Morf, M., Sidhu, G. S., and Kailath, T., 1974, Some new algorithms for recursive estimation in constant linear discrete-time systems. *I.E.E.E. Trans. autom. Control*, **19** (4), 315–323.

Morf, M., Vieira, A., and Kailath, T., 1978, Covariance characterization by partial autocorrelation matrices. *Annals of Stat.*, **6** (3), 643–648.

Morf, M., Vieira, A., and Lee, D., 1977, Ladder forms for identification and speech processing. *Proc. I.E.E.E. Conf. Decision Control* (New Orleans, LA), pp. 1074–1078.

Potter, J. E., 1966, Matrix quadratic solutions. *SIAM J. Appl. Math.*, **14,** 496–501.

Newkirk, J., 1979, Computational issues in linear least-squares estimation and control. Ph.D. Dissertation (Dept. of Elect. Engng, Stanford University, Stanford, CA).

RISSANEN, J., and BARBOSA, L., 1969, Properties of infinite covariance matrices and stability of optimum predictors. *Inform. Sci.*, **1,** 221–236.

ROBINSON, E., 1967, *Multichannel Time Series Analysis with Digital Computer Programs* (San Francisco, CA : Holden-Day).

RODRIGUEZ-CANABAL, J. M., 1973, The geometry of the Riccati equation. Ph.D. Dissertation (USC, Los Angeles, CA). Also *Stochastics*, **1** (2), 129–149.

SEBER, G., 1977, *Linear Regression Analysis* (New York : J. Wiley & Sons).

SIDHU, G. S., 1975, A shift-invariance approach to fast estimation algorithms. Ph.D. Dissertation (Dept. of Elect. Engng, Stanford University, Stanford, CA).

STEWART, G. W., 1973, *Introduction to Matrix Computations* (New York : Academic Press).

VERGHESE, G., FRIEDLANDER, B., and KAILATH, T., 1977, Scattering theory and linear least-squares estimation, Pt. III : The estimates. *Proc. I.E.E.E. Conf. Decision Control* (New Orleans, LA), pp. 591–597. Also *I.E.E.E. Trans. autom. Control*, (1980), **25** (3), 000–000.

WHITTLE, P., 1963, On the fitting of multivariate autoregressions and the approximate canonical factorization of a spectral density matrix. *Biometrika*, **50,** 129–134.

WIGGINS, R. A., and ROBINSON, E., 1965, Recursive solution to the multichannel filtering problem. *J. Geophys. Res.*, **70,** 1885–1891.

WILLEMS, J., 1971, Least squares stationary optimal control and the algebraic Riccati equation. *I.E.E.E. Trans. autom. Control*, **16,** 621–634.